CAMBRIDGE LIBRARY COLLECTION

Books of enduring scholarly value

Technology

The focus of this series is engineering, broadly construed. It covers technological innovation from a range of periods and cultures, but centres on the technological achievements of the industrial era in the West, particularly in the nineteenth century, as understood by their contemporaries. Infrastructure is one major focus, covering the building of railways and canals, bridges and tunnels, land drainage, the laying of submarine cables, and the construction of docks and lighthouses. Other key topics include developments in industrial and manufacturing fields such as mining technology, the production of iron and steel, the use of steam power, and chemical processes such as photography and textile dyes.

Present Day Metallurgical Engineering on the Rand

The discovery of gold in the Witwatersrand hills and the Transvaal region of South Africa sparked a rush in the late 1880s. Competition between the British and Boer settlers for access and control of this valuable commodity was one of the underlying causes of the second Anglo-Boer war (1899–1902) in which the British eventually won control of the territory. In this work, published in 1898, the mining engineer and Fellow of the Geological Society of London John Yates outlines the state of the booming industry on the eve of war. He discusses the work of key companies and includes technical specifications and illustrations of the equipment used in the new gold mines, such as the shafts, mills and cyanide works. An appendix by his fellow engineer Hennen Jennings addresses the question of government assistance in subsidising the huge start-up costs of these mining projects.

Cambridge University Press has long been a pioneer in the reissuing of out-of-print titles from its own backlist, producing digital reprints of books that are still sought after by scholars and students but could not be reprinted economically using traditional technology. The Cambridge Library Collection extends this activity to a wider range of books which are still of importance to researchers and professionals, either for the source material they contain, or as landmarks in the history of their academic discipline.

Drawing from the world-renowned collections in the Cambridge University Library, and guided by the advice of experts in each subject area, Cambridge University Press is using state-of-the-art scanning machines in its own Printing House to capture the content of each book selected for inclusion. The files are processed to give a consistently clear, crisp image, and the books finished to the high quality standard for which the Press is recognised around the world. The latest print-on-demand technology ensures that the books will remain available indefinitely, and that orders for single or multiple copies can quickly be supplied.

The Cambridge Library Collection will bring back to life books of enduring scholarly value (including out-of-copyright works originally issued by other publishers) across a wide range of disciplines in the humanities and social sciences and in science and technology.

Present Day Metallurgical Engineering on the Rand

JOHN YATES

CAMBRIDGE
UNIVERSITY PRESS

One 1,000 I.H.P. Horizontal, Compound, Gates-Bates-Corliss Engine, with 21-in. and 42-in. Cylinders and 60-in. stroke.
One 500 I.H.P. Auxiliary Engine by Ingersoll-Sergeant : Compound, Horizontal, with Corliss valve gear ; Cylinders, 18-in. and 36-in.
     diameter, stroke, 44 in.
One, 16 drill capacity, Ingersoll-Sergeant Horizontal Compound Compressor.
One, 42 drill capacity, Philadelphia Corliss Compound Horizontal Compressor, by Philadelphia Engineering Company.
Two Elwell-Parker Generators, Volts, 110, Amps., 220, Revs., 730.   For the pumps of the Cyanide Works and for lighting.

GINE ROOM

AINING :

One Easton, Anderson & Goolden, Limited, Generator, Volts, 105-110, Amps., 455, Revs., 750, for lighting.
One Easton, Anderson & Goolden, Limited, Generator, Volts, 610, Amps., 82, Revs., 760, for pumping.
One Easton, Anderson & Goolden, Limited, Generator, Volts, 220, Amps., 364, Revs., 625, for Cyanide Works.
Two Easton, Anderson & Goolden, Limited, Generators, Volts, 585, Amps., 144, Revs., 620, for mine pump.
One 40 horse-power three-phase Generator of 2,000 Volts, by Baden, Brown, Doveriei & Company.

PRESENT DAY
METALLURGICAL ENGINEERING
__ ON THE RAND __

BY

JOHN YATES,

(ASSOCIATE OF THE ROYAL SCHOOL OF MINES, LONDON; FELLOW OF THE
GEOLOGICAL SOCIETY OF LONDON; MEMBER OF THE FEDERATED
INSTITUTE OF MINING ENGINEERS.)

WITH AN

APPENDIX

ON THE

ECONOMICS OF THE TRANSVAAL GOLD
MINING INDUSTRY

BY

HENNEN JENNINGS,

(CONSULTING ENGINEER TO MESSRS. ECKSTEIN & CO.)

LONDON:
OFFICE OF "THE MINING JOURNAL,"
18 FINCH LANE, E.C.

1898.

INDEX.

———•—————

INDEX.

ILLUSTRATIONS.

———•———

PREFACE.

WHEN the writing of this small volume was undertaken, it was not intended that its scope should extend beyond description of the Engineering connected with the gold industry of the Witwatersrand.

The universal interest, however, which has been aroused by the prevailing depression on these Fields, and the general sympathy with the mines which has followed the publication of their grievances has induced me to give, as an appendix, a statement made by Mr. Hennen Jennings before the recent Mines Commission appointed to investigate and report on the present state of the industry. This statement—one of several admirable ones made before the Commission—is a clear, reliable, and comprehensive review of the economics of Rand Mining, and will repay a careful study by all concerned. It discloses the burdens under which the mines are struggling, and it brings forward figures which testify that, notwithstanding the tardiness of the Transvaal Government to encourage an industry which is vital to the prosperity of the State, the mines have adopted a liberal policy and done everything in their power to develop the industry by conducting it on sound lines.

We are now at an important period in the history of the Rand and the State, for the time has arrived when our able Engineers and Managers can improve the working of the mines but little further; and as it has been clearly demonstrated that under existing conditions very few properties can work at a profit, it follows that, unless the Government affords the much needed relief from present burdens, the brilliant future which has so often been predicted for these Fields will never be attained.

JOHN YATES.

JOHANNESBURG, *January*, 1898.

PRESENT DAY METALLURGICAL ENGINEERING ON THE RAND

INTRODUCTION.

THE world presents no greater display of mining and metallurgical enterprise than lies within view of Johannesburg. To-day the Witwatersrand as a gold field stands unique in the number and magnitude of its mines, for though, elsewhere, there are mines which surpass in the complexity of their equipments or in their scale of operations the average Witwatersrand venture, there is no other field which discloses such a continuous run of large and well found mines. The average standard of equipment of the whole of the Rand properties is high, and the financial results arising from this standard reflect great credit upon Rand engineers who, by their ability and enterprise, are fast making Johannesburg the hub of their profession. In their hands both mining and metallurgy have made great strides, and that desideratum— 100 per cent. extraction on 100 per cent. of the ore—is being brought very close indeed, whilst as regards working costs these compare favourably with those of other fields, notwithstanding the burdens and disadvantages under which the gold industry of the Transvaal labours.

Fortunately the Rand ores are not of a complex character, but are amenable to treatment of a comparatively simple nature, and to this fact, and to the exceptional nature of the reefs— the unprecedented continuity both of reefs and gold contents—the industry owes its enormous development, if not its existence, and the world a great source of gold, for the ore is of such comparatively low grade that had it proved refractory, or the reefs of uncertain character, few properties could have been opened with reasonable prospects of developing into payable propositions. Even as it is, with all the advantages accruing from the exceptional nature of the reefs, the majority of the Rand properties have such poor ore that it is only by the strenuous efforts of the engineers—their striving after perfection in method, plant, and process—that some few of them are able to show a margin of profit, or others ever expect one under present economic conditions.

Whilst, however, from the industry's point of view it is regrettable that the average gold contents of the banket is small, and that the economic conditions are unfavourable, these adverse factors have undoubtedly contributed materially to the advance of metallurgical engineering on these fields by creating a demand for the high skill required to so meet these conditions as to work with commercial success, a demand which has attracted to Johannesburg many of the leaders of the mining profession. In their efforts in this direction engineers have been ably assisted by directors, who have not hampered them by limitation of funds, but given them a free hand and permitted a liberal expenditure, wisely recognising that capital judiciously spent in installing the best of plant, and applying the most approved methods, is likely to be more than returned by low working costs.

It is not surprising that under these conditions engineering on the Rand should have attained a high degree of perfection, and that its course should be watched with keen interest by all connected with gold mining; for in these days, when the possibilities attending the vast potential wealth lying in the low grade reefs of the world are beginning to be realised, the industry recognises the importance of any improvement giving even $\frac{1}{2}$ dwt. extra profit per ton, hence descriptions of such are welcome.

Of course, it is hardly necessary to mention that the mining and metallurgical practice suited to one field may not be adapted to the conditions obtaining on another, but even when a difference of conditions exists, a knowledge of how the difficulties in one country were surmounted is invariably of assistance in solving the problems pertaining to others.

Thus the comparatively simple metallurgical practice of the Rand would not be applicable, at least in its entirety, on many other gold fields; but, nevertheless, a description of present-day Rand practice—the outcome of experience which has cost millions of pounds —cannot fail to be of value. Now and again general accounts of Witwatersrand engineering have been put forth, but so rapidly are things moving on these fields that few, if any, correctly depict present-day practice. So far no attempt has been made to describe the surface installations of the mines in detail, and it is to meet this lack of description of constructive details and to record the latest developments—more particularly in connection with mill and cyanide engineering—that I have written the following pages. In them I give the arrangements at three well-known mines for handling and treating the ore in its course from the skip to the residue dumps, the mines selected as exemplifying the latest ideas and tendencies being the Glencairn Main Reef Gold Mining Co., Limited (160 stamps), an outcrop property of the Barnato group of mines, the Simmer and Jack Proprietary Mines, Limited, of the Consolidated Gold Fields group (280 stamps, the largest gold mill and cyanide works in the world), also an outcrop company, and the Jumpers Deep, Limited (200 stamps) of the Rand Mines (Limited), a deep level property, the surface works of each of these three properties being more or less typical of its kind, and possessed of distinct features of interest. The mills of these three plants are not yet running,* being still in course of construction, but those of the Glencairn Main Reef and the Simmer and Jack are approaching completion.

This description I supplement by a general review of Rand practice.

The treatment of mill slimes has been receiving attention on the Rand for several years, and many methods have been tried with more or less success, but it is only within the past few months that a process has been developed free of the objections attached to its fore-runners, and promising commercial success on the average slime of the Witwatersrand. In working at this slime question the difficulties met have been almost entirely of a mechanical nature. The gold in fresh slime is readily dissolved by cyanide, but to successfully apply this solvent on a working scale has proved a severe tax on ingenuity.

The method now being successfully worked consists, in the main, of agitating the slime with very dilute solutions of cyanide of potassium, and then settling and decanting the clear auriferous solution, precipitating the gold from it by electricity.

There is no questioning that the solution of the slime problem marks an era in metallurgical science. Whilst not of such vital import to the welfare of the Transvaal gold industry as the introduction of the cyanide process, its importance to this country can be realised when it is remembered that about 1,200,000 tons of slimes were produced on the Rand in 1896, which, if an average gold contents of 3½ dwts. per ton be assumed, carried precious metal to the extent of about £840,000. Treated at a cost of 5s. per ton, these slimes would have given a profit on every ton of (approximately) 5s., assuming a 70 per cent. extraction, or a total profit of £300,000, equal to a 5 per cent. dividend on £6,000,000.

These considerations are sufficient to clothe the three pioneer plants at work—the Robinson the Crown Reef, and the Rand Central Ore Reduction Companies—with much interest, and the description of the Crown Reef and of the Jumpers Deep slime plant which I give is additionally interesting, inasmuch as the former is doing very successful work, and the latter is of the latest design, its construction having only just been put in hand.

* All three plants are now running.

DESCRIPTION OF PLANTS.

THE GLENCAIRN MAIN REEF GOLD MINING CO., LIMITED.

THIS property, consisting of 144 claims, is under the general management of Mr. J. Blane, also general manager of the New Primrose Mine (160 stamps), and one of the most able managers on the Rand.

It is worked by two main inclined shafts: one, the Glencairn shaft, being 280 feet from the mill; and the other, the Glenluce shaft, is 1,800 feet from the latter, with which both are connected by endless rope haulage.

The Glencairn main shaft has two skip ways, the skips, which are of $\frac{1}{2}$ in. plate, being of 3 tons capacity and automatic tipping. The height from the headgear sills to the centre of the winding sheaves is 42 ft., the gear legs being of 12 by 12 in. timbers resting on 16 by 12 in. sills, with framing and ties of 12 by 12 in. The shaft is laid with 9 by 9 in. runners carrying 45 lb. rails, the angle of dip being approximately 37°. The skips discharge on to a grizzly 10 ft. by 9 ft. 6 in., constructed of bars $2\frac{1}{2}$ by $\frac{3}{4}$ in., placed with $1\frac{1}{4}$ in. spaces, and at an inclination of 35°. The "deads" are tipped into a special bin. The coarse ore rolls from the grizzly to a washing floor, 11 ft. by 5 ft., covered with $\frac{1}{2}$ in. iron plates. Here a jet of water is played on it, and it is then scraped on to a travelling belt for sorting and conveying to the crushers, the waste taken out being cast into a bin alongside of which the belt travels. This belt is 36 ft. between the centres of the end pulleys and 2 ft. 9 in. wide with a travel of 38 ft. per minute and a rise of 6 in. to the crusher floor, where it tips the rock between two crushers. It is constructed of $\frac{1}{4}$ in. iron plates, these being the width of the belt and 1 ft. $7\frac{1}{2}$ in. long, with other $\frac{1}{4}$ in. plates rivetted to them to take the wear. The several plates are linked together with $1\frac{1}{4}$ in. pins, and overlap each other $1\frac{1}{2}$ in., the belt being supported and running on 12 in. wheels.

The crusher floor is 20 ft. 6 in. by 14 ft. 8 in., and is laid with $\frac{1}{4}$ in. iron plates. The crushers are two No. 5 Gates, and from these the ore falls to the bin beneath, by which runs the tramway to the mill. The power for the whole of the machinery in this headgear is taken from a 4 in. shaft driven by a 16 H.P. non–condensing horizontal Ruston and Proctor engine running at 100 revolutions per minute, whilst the winding engine is a 20 H.P. non-condensing horizontal geared Hornsby. This latter, however, is to be replaced by a direct-acting Corliss valve engine 18 by 48 in., with 8 ft. drums.

The Glenluce main shaft has three skipways, one being reserved for development work. The skips are similar to those at the Glencairn shaft, but that for development is of 2 tons capacity. The headgear is 63 ft. from the collar to the centre of the sheaves (8 ft. diameter), the legs being of 14 by 14 in. timber strongly framed and braced with 12 by 12 in. and 12 by 6 in. timber and $1\frac{1}{4}$ in. bolts, the sills being of 14 by 7 in. section. The six runners are 12 by 8 in., and carry 45 lb. rails, the angle of the shaft at the surface being 60°.

The grizzly in this headgear is 14 ft. 6 in. by 13 ft. 6 in., set at an angle of 35°, and of a similar construction to that at the Glencairn shaft. To reach the grizzly the ore slides over a sloping door opening to the waste bin, and from the bars it falls upon an inclined shaking table 12 ft. long and 3 ft. wide, set with an inclination of 7°, and making 150 strokes per minute in the direction of its length, being driven by an eccentric and rod. This feeds the ore to a revolving circular sorting table, consisting of an annular ring 2 ft. 6 in. wide and 24 ft. outside diameter, with a slight down slope to its outer edge. The table, which is supported by a 6 in. central vertical shaft, is driven at three-quarters of a revolution per minute by two wood conical friction pulleys, 2 ft. diameter and 9 in. face, these latter being in contact with the underside of the table at diametrically opposite points. On this table the ore is washed and the waste sorted out and thrown into a bin below the centre of the table, whilst the ore is carried round to two Wells' patent crushers, described hereafter, and falls thence to the rectangular ore bin, 38 ft. by 28 ft. by 10 ft. deep, into which the fines passing the grizzly also gravitate, and beneath which are two tram tracks for the trucks conveying the ore to the mill. The sorting and crushing machinery at this shaft, together with the endless haulage to mill, are driven by 50 H.P. Hornsby tandem condensing engine running at 60 revolutions and driving a 5 in main shaft at 160 per minute.

The winding engines here are two in number, one a 30 H.P. Worsley-Mesnes direct-acting non-condensing horizontal for the development work, and the other a 40 H.P. Ruston Proctor geared, non-condensing. Both engines have double drums and are so arranged that either engine can take two of the skips.

The endless rope to the mill is ⅞ in. diameter and travels 1½ miles per hour. The highest gradient is on the trestle leading to the top of the mill ore bin, the grade here being 1 in 8. The trucks are side-tipping and of 20 cubic feet capacity and 18 in. gauge, the rope riding them and catching by the swivel jockey arrangement in general use on the Rand.

THE MILL.

The mill contains 160 stamps of 1,250 lbs. Sandycroft foundry make, arranged back to back in two rows of 80, with an ore bin between them. The mill building is 178 ft. long by 76 ft. wide, and from its floor to the top of the ore bin measures 34 ft.

The ore bin is 151 ft. long by 19 ft. wide, the depth on the centre line being 12 ft. 6 in., and at the sides 21 ft. 6 in., the bottom being A-shaped. The foundation posts of the bin rest on the line sills common to each row of stamps, which in turn bear on nine mud sills, five of 12 by 12 in. section, and four of 14 by 14 in. ; these larger timbers running one at the front and one at the back of the mortar blocks with a 5½ in. space. Of the five 12 by 12 in. mud sills there is one in front of each line of stamps, 3 ft. from the front 14 by 14 in. mud sill, and the others are under the ore bin foundation posts—one at each side of the bin and one in the centre. All the mud sills are bedded on hard tamped ground. The line sills resting on and crossing the mud sills are all 18 in. deep, those supporting the centre battery posts—the stamps being arranged in batteries of ten with a 5 ft. 7 in. space between them — being 18 in. broad and the two intermediate ones between these 12 in. broad. From centre to centre of the line sills is 6 ft. 6 in. The three ore bin foundation posts resting on each of these sills are of 12 by 12 in. section with two 12 by 12 in. braces running to the top of the centre one. On these posts rest the ore bin sills, 14 by 14 in., the ends of which form the battery knee posts, and each of these carries the three bin posts, 12 by 12 in., one at the centre and one at each side of the bin tied together by a 9 by 9 in. timber at the top and a 12 by 12 in. at the lower level, both passing across the bin.

The bottom of the bin rests on 12 by 12 in. inclined timbers, stiffened at their centre by 12 by 12 in. struts. The sides and bottom of the bin are built of 3 by 9 in. deals, those of

C

the bottom being supported by 3½ by 5½ in. deals 2 ft. apart, checked into the 12 by 12 in. inclined timbers. The bolts used throughout the bin are 1⅛ in. and 1½ in. diameter, and iron plates ⅜ in. thick line the whole of the bin interior. Over the bin, which contains no partitions, run two tram tracks with the endless haulage connecting with the shafts.

The stamps, arranged as before described, have a double space between the 50th and 51st stamps of each row, the clean-up pans being located in the passage thus formed.

The mortar block of each battery of five stamps is built up of eight 15½ by 15½ in. baulks 18 ft. long, these being bolted together with 1 in. bolts and having an 8 by 8 in. timber key running through them. Some of these piles are bedded on solid rock and others on concrete, and around them is hard tamped earth. Their upper end, which projects 4 ft. above the level of the line sills, is kept in position by two 7½ by 7½ in. timbers, one at the back and one at the front, these being bolted to the battery posts. The latter are 21 ft. 10 in. high from the line sills, the two side ones of each battery of ten stamps having a 24 by 12 in. section, and the centre one 24 by 18 in. They are each braced to the ore bin sill, which butts against them—forming the knee beam—1 ft. 7 in. below the level of the cam shaft centre and 9 ft. from the top of the posts, and to a post of the ore bin by a 12 by 12 in. horizontal timber 19½ in. from the top of the battery posts. The tie bolts are 1¼ in. diameter. A bolt running diagonally secures the bottom of the battery posts to the line sills, whilst laterally they are supported by a cap sill 12 by 12 in. running the length of the series, and the two guide beams, 12 by 14 in., ⅞ in. bolts binding these to the battery posts. The back cam floor, 5 ft. 8 in. wide, is carried by the knee beam, whilst the front floor at the same level is borne on deals which have one end secured to the king posts and the other suspended by ⅞ in. bolts from the tie beam of the roof truss. The upper and lower guide beams are 3 ft. 2 in. and 11 ft. 8 in. from the top of the posts respectively, and from the centre of the cam shaft to the line sills is 14 ft. 5 in. All the mill timber is of Oregon pine.

The mortars are 4 ft. 10 in. long by 19 in. broad (at the top) by 5 ft. 7 in. deep—all external dimensions—and weigh about 4 tons. Some of them are bedded on tarred blanketing, others on ¼ in. sheet rubber. The thickness of the sides of the mortar at the top is 1¼ in., and the base of the mortar is 10½ in. thick carrying a flange 4 in. thick projecting 5 in. for the anchor bolts, these latter, eight in number, being 1½ in. diameter and 3 ft. 6 in. long, the lower end being secured by cotter and washer.

The discharge aperture is 22 in. deep and the feed inlet 2 ft. long by 4¾ in. wide its lower lip being 19¼ in. above the bottom of the mortar. The inner edge of the mortar

Glencairn Main Reef Gold Mining Company, Limited.

View of Battery.

discharge lip is $10\frac{1}{4}$ in. above the bottom, and the breadth of the mortar at this level is $14\frac{1}{2}$ in. to the screen frame, the latter being set at an angle of 10° from the vertical. These dimensions, however, are not the working size of the mortar inasmuch as there are $\frac{3}{4}$ in. steel liners on the ends and sides, the former being 20 in. deep and the latter $10\frac{1}{4}$ in. deep, whilst a cast-iron plate $2\frac{1}{4}$ in. thick—in two sections—forms a false bottom. The screening is 700 mesh, and the frame presents one aperture 4 ft. by $9\frac{3}{4}$ in., the framing being $2\frac{1}{2}$ in. deep and 2 in. thick. On the screen frame rests a 2 in. board which closes the aperture of the mortar above the screen.

The mortars have an amalgamated plate $\frac{1}{8}$ in. thick and 11 in. deep fixed just above the liner at the back.

The dies, which have their centres 10 in. apart, are of forged steel and weigh 130 lbs. They have a cylindrical body and a square base with bevelled corners; the former is 9 in. diameter and 5 in. deep, and the latter $9\frac{1}{2}$ in. square and $1\frac{1}{2}$ in. deep.

The steel stamp stems are 15 ft. long and $3\frac{1}{2}$ in. diameter, a taper 5 in. long on each end reducing this to $3\frac{5}{16}$ in. They weigh 490 lbs. The cast steel heads, weighing 376 lbs., are 2 ft. high by 9 in. diameter, whilst the shoes of forged steel, weighing 240 lbs., have butts 12 in. high and 9 in. diameter, the shanks being $5\frac{1}{4}$ in. high and $4\frac{1}{2}$ in. diameter tapering to $3\frac{1}{2}$ in.

Tappets of the usual Californian gib pattern in cast steel are used. They are 12 in. deep by $8\frac{1}{2}$ in. diameter, presenting a $2\frac{1}{4}$ in. face. They weigh 144 lbs. and the gib is wedged up by two keys.

Each five stamps is driven by a separate 6 in. cam shaft having its centre 6 in. from that of the stem and carrying a wood pulley 6 ft. diameter and 12 in. face with a 10 in. Dick's belt. The cam shaft bearings on the side posts are 12 in. long and the centre double bearing 20 in. long. The cams, double-toed and of cast steel, are 33 in. across the toes and have $2\frac{1}{2}$ in. faces. Their hubs are 11 in. diameter and 5 in. long and are secured to the shaft by Blanton's patent key, the side of the cams being set to clear the stems by $\frac{3}{16}$ in. The stamps will make about ninety 9 in. drops per minute, the order of falling being one, three, five, two, four.

The guides are of Blane and Wells' special pattern, consisting of cylinders of karri wood 16 in. long, bored to take the stem, the thickness of the cylinders formed being $1\frac{1}{8}$ in. These cylinders or collars, which are in halves, are held in neat castings each bolted to the guide beams by two $\frac{5}{8}$ in. bolts, eight of these passing through a strapping plate 3 ft. $3\frac{1}{2}$ in. by

6 in. by $\frac{1}{2}$ in. on each of the guide beams. A $3\frac{1}{2}$ in. Jack shaft carries the finger bars, and a crawl with a 30 cwt. chain block runs on two 9 in. by 3 in. deals bolted to the roof tie beams.

The amalgamated copper tables, which are 3 in. below the outer lip of the mortar, each consists of one plate 11 ft. 10 in. long by 4 ft. 9 in. wide and $\frac{1}{8}$ in. thick, set with a fall of 1 in 10. Between it and the table is one thickness of blanket, and the plate is secured at top and bottom by screws and at the sides is bent up under a wood fillet. The table at the top end bears on a deal carried by the $7\frac{1}{2}$ in. by $7\frac{1}{2}$ in. timber supporting the mortar block, and at the bottom end rests on a light trestle.

Across the bottom of the table is a launder 6 in. wide sloping towards a vertical central $2\frac{1}{2}$ in. pipe which drops the pulp into a mercury trap, this taking the form of a rectangular box 2 ft. 3 in. long, 12 in. broad, and 12 in. deep divided into four parts by three copper plate vertical partitions which are so arranged as to make the pulp passing through the box flow upwards and downwards alternately. From the lip of this trap the pulp passes to cement culverts leading to the tailings pumps. The floor of the mill has a slope to the sides and the height of the bottom end of the amalgamated tables above it is 23 in.

The ore is fed into the mortars by suspended challenge feeders and the water supply for the mortars comes from a reservoir in a 10 in. main to the condensers, and thence, after warming, passes by two 8 in. pipes along the mill. From these latter a $1\frac{1}{2}$ in. branch rises in front of each side battery post and discharges into the mortar, whilst a 3 in. branch rises in front of each centre battery post and divides into two $1\frac{1}{2}$ in. pipes, one going to each box, each mortar thus having two $1\frac{1}{2}$ in. service pipes.

The clean-up apparatus consists of two iron grinding pans 3 ft. diameter and 10 in. deep, by Harvey of Cornwall, an amalgamated table 8 ft. long by 2 ft. wide, and one amalgam retort.

The mill is driven by a horizontal, cross-coupled, compound condensing Gates-Bates-Corliss engine, with cylinders 21 in. and 42 in. diameter and a stroke of 60 in. Its driving wheel is 23 ft. diameter and carries thirty $1\frac{1}{2}$ in. diameter cotton ropes. These drive a 10 in. countershaft at 96 revolutions, from which the two lay shafts of the mill are driven at the same speed; these shafts, which have their bearings on the line sills behind the battery posts, being 7 in. diameter at the one end and 4 in. at the other. The cam shaft belts are thrown into action by means of belt tighteners, the tightener pulley being 16 in. diameter.

This main engine also drives the dynamos which furnish the power and light needed about the surface works.

(25)

Glencairn Main Reef Gold Mining Company, Limited.

View of Cyanide Works, showing Intermediate and Leaching Vats, and the Extractor House and Sump Vats.

The auxiliary engine is an Ingersoll-Sergeant of the compound, horizontal, cross-coupled, condensing type, with Corliss valve gear. Its cylinders are 18 in. and 36 in. diameter and the stroke 44 in., and its 16 ft. fly-wheel carries twenty-four cotton ropes of 1½ in. diameter.

THE CYANIDE PLANT.

The cement culverts carrying the pulp from the mill convey it to the sump of the tailings pumps, these being situated in an annexe of the mill building. There are two 15 in. double plunger pumps with a stroke of 4 ft., each plunger making 16·8 throws per minute, the lift being about 45 ft. Each pair of plungers is driven by cranks and connecting rods on a separate 7 in. shaft, the power being conveyed from the mill to each pair by a 9 in. belt which runs on a 4 ft. by 10 in. pulley on the pump countershaft; this shaft—which is 5 in. diameter and makes 84 revolutions per minute—having a 13¼ in. pinion gearing with a 5 ft. 6¼ in. spur wheel on the pump crank shaft. Each pair of pumps is calculated to be equal to lifting the total output of the mill and one pair only will be run. The four suction and rising mains—one for each cylinder—are 10 in. diam., each pump having an independent delivery main. The pumps, which are of simple construction and were made by Mitcheson, Durban, are packed with tallowed hemp, and the valves, which are 1 in. rubber discs, rest on cast iron seats.

The pumps are carried by brackets resting on 14 by 22 in. timbers on masonry foundations, the crank shaft bearings being on 16 by 16 in. sills carried by 14 by 14 in. posts, these latter standing on 14 by 14 in. sills securely anchored to masonry. The sump bottom is cemented and has a wash out pipe. The four rising mains deliver into a launder 15 in. wide which has a 4 per cent. grade to the spitzluten, three in number. In the spitzluten a concentrate is separated from the pulp, a by-pass being provided by means of which the boxes can be thrown out of action if desired. The spitzluten are all 3 ft. 6 in. deep, and 2 ft. wide, but of varying lengths, the first being 3 ft. 6 in. long, the second 3 ft. 9 in., and the third 4 ft. The long sides of each box are vertical to a depth of 22 in., but from there have an inward inclination of 20°. The other sides have an inward inclination of 30° in the first box but somewhat less in the other two, the four sides of each box thus forming an inverted truncated pyramid, the truncated area—the base of the spitzluten—being 8 in. by 8 in. Each base has a 2½ in. discharge hole connected with a 2½ in. water pipe and a discharge nozzle which in the first box is 1¼ in. diameter and in the second and third 1 in. diameter. The boxes, which are built of 2 in. deals with 3 in. by 2¼ in. posts and ⅝ in. bolts, have each a vertical baffle board. The launder from the spitzluten fans out to 4 ft. wide for the introduction of

D

an automatic sampler. This latter has been designed by Mr. MacBride, one of the leading cyanide managers on the Rand, who is in charge of these works. It consists of a horizontal pipe of pear-shaped section, its narrow end, which is the upper one, having a ⅝ in. slot. This pipe is supported parallel to the lip of the launder and under it by two vertical posts pivoted on the sides of the latter so that the pipe can be carried through the issuing stream of pulp. The sampler is actuated by two rods connected with a V-shaped tilting water trough which is divided longitudinally by a partition. This trough tilts on both sides of the vertical, giving the sampling pipe two positions of rest—one below the stream of pulp and the other above it, the samples being thus taken with the pipe ascending and descending alternately. Springs restrain the motion of the tilting trough.

Leaving the fan-shaped launder from the spitzluten the battery pulp passes into another launder, 8½ in. by 11 in. deep, which runs the length of the plant and has branches to the distributor of each intermediate vat.

The cyanide vats are arranged in two tiers, one directly above the other, each tier comprising two parallel rows of seven vats each, the long axis of the plant running parallel with that of the mill, and the top of the mortar blocks being level with the rims of the lower series of vats. The plant, which has a capacity of about 17,000 tons per month, allowing for about 4½ days' treatment, has cost £28,000 and consists of 28 vats—14 intermediate vats 30 ft. diameter by 7 ft. 3 in. deep, and 14 leaching vats 30 ft. by 8 ft. deep, all internal dimensions. Of these, two intermediate vats and two leaching vats are reserved for concentrates. The intermediate vats are built of selected Baltic deal, the staves being 8 ft. long by 8 in. broad by 2¾ in. thick, and the bottoms of 9 in. by 3 in. deals with dowels. Each vat is bound by eight 1⅛ in. iron hoops, the several lengths of each hoop being connected by cast-iron junction boxes. The vats are each provided with four bottom discharge doors, having 15 in. apertures, through which the charge is thrown into the leaching vat beneath. These doors open from below, access being given by a light platform over the lower vats. Some of these intermediate vats are fitted with ordinary filter bottoms and the others with a special filter consisting of four rectangular frames covered with ½ in. mesh wire screen and two filter cloths, the frames, which are 12 ft. long by 5 ft. 6 in. broad, being laid in cement. For introducing and distributing the battery pulp all but the two concentrate vats have rotating distributors these consisting of a central conical bowl free to rotate on a vertical iron column, which bowl has radiating from it a number of pipes of various lengths the outer free ends of these being so bent that the reaction due to the issuing pulp suffices to make the cone with its pipes rotate and so effect a distribution of the pulp.

The water and slime escape through two slat gates fixed to the side of the vat. These gates each consist of two vertical cast-iron posts, placed 1 ft. 9 in. apart, which run from the top of the vat to the level of the filter where they rest on a timber base, 2 ft. 8 in. long by 3 in. thick, projecting 9 in. from the side of the vat and placed 2½ in. from the bottom. These posts present vertical grooves 1½ in. wide, down which wood slats, 1 ft. 9 in. by 4½ in. deep and 1¾ in. thick reduced to 1⅜ in. at the edges, are dropped as the filling of the vats proceeds, the distance of the slats from the side of the vat being 4½ in. on the centre line. The water and slime flow over these slats and find exit through an 8 in. hinged lid discharge valve fixed to the side of the vat near the bottom. To render the joint between the slats and the cast-iron posts as sand-tight as possible a strip of rubber, 3½ in. wide and ⅛ in. thick, running the full length of the posts, is so secured to each of the latter as to overlap the slats; the sands deposited in the vat pressing on this tends to form a tight joint. The overflow valve discharges into a launder which conveys the stream to a spitzkasten. In the intermediate concentrate vats the launder from the bottom of the three spitzluten delivers into a hexagonal box fixed in the centre of each vat to act the part of a distributor; the box, which is 2 ft. across the flats and 18 in. deep, having a number of holes in the sides and at the bottom.

The leaching vats proper are of iron and are 30 ft. diameter by 8 ft. deep. The sides, which are one plate deep, are of $\frac{3}{16}$ in. iron with single rivetted lap joints, the rivets being ½ in. diameter, the pitch 1⅝ in., and the overlap ⅞ in. The rim of the vat is strengthened by an external angle iron 2 in. by 2 in. by ¼ in. secured by ½ in. rivets of 5 in. pitch, whilst an external channel iron stiffener, 4 in. by 1½ in. by $\frac{3}{16}$ in., encircles the vat at mid-depth. An internal angle iron, 2 by 2 in. by ¼ in., secures the bottom to the sides of the vat, the plates of the bottom, which are ¼ in. thick with radial joints, being rivetted to this angle iron on their outer edge and to a circular centre plate 4 ft. diameter on their inner edge. The joints are single-rivetted lap joints and each bottom is fitted with six circular discharge doors, 15 in. diameter, of MacBride and Brown's patent design. The filter frame in these vats each consists of 1⅛ in. by ⅞ in. slats fixed with 1¼ in. spaces across 4¼ by 1½ in. boards placed with centres 12 in. apart and standing on their 1½ in. faces. The underside of these latter boards has circular cuts taken out here and there to permit free passage of the solution. Around the outer edge of this filter frame, which is made in segments to facilitate its being taken up, a board 4¼ in. deep by 1½ in. is secured which comes within 1¼ in. of the vat side, this space being taken up by the lagging rope which holds down the filter cloths. The latter are two in number, a jute cloth below with a coir matting resting on it. Over the coir are 2 by 1⅛ in. slats, placed 6 in. apart, to protect the cloths from the shovels.

These iron leaching vats are coated internally with tar, and externally are painted grey like the intermediate vats and plant generally.

The twenty-eight vats, which are not housed, are carried by stout timbers resting on seven masonry piers running the length of the plant and forming six passages—three beneath each row of lower vats—for the discharge trucks. These piers are built of large roughly squared stone set in hydraulic lime mortar, and are 2 ft. 6in. thick at the top and 3 ft. 6 in. at the bottom or ground level, the centre pier—the one between the two rows of vats—being somewhat larger, viz., 3 ft. thick at the top and 4 ft. at the bottom. From the top of the piers to their footings is 5 ft., the latter being of masonry which projects 6 in. on each side of the piers and goes down about 6 ft. to firm ground. The two centre passages formed by the seven piers are 8 ft. 9 in. wide at the top, whilst the two intermediate ones are 6 ft. 6 in. and the two side ones 7 ft. 1½ in.

On each of the masonry piers rests a wall plate consisting of two 12 by 6 in. timbers lying flat and forming a 24 in. bearing for the transverse joists of which those under the lower vats are 14 in. deep but of various breadths. At each end of the series of vats and between each pair of the iron vats—the centres of the vats being 33 ft. 3 in. apart the long way of the plant and 32 ft. the other way—these joists are 12 in. deep and 14 in. broad, whilst under each vat are three joists 14 by 14 in. carrying 14 by 14 in. posts. Between the 14 by 14 in. transverse joist under the centre of the vat and the similar one on each side of it are two 8 by 14 in. joists, whilst beyond these each end of the vat is supported by two 6 by 14 in. joists. On these transverse timbers the iron vats directly rest, each of these vats thus having under it three 14 by 14 in. joists, four 8 by 14 in. joists, and four 6 by 14 in. joists. Upon the 14 by 12 in. and 14 by 14 in. joists, at every point where they bear on the wall plates, stand posts of 14 by 14 in. section, seven standing in each vat and thirteen around it, eleven of the latter being common to the adjoining vats except in the case of the end vats where only eight are common. The odd number of posts standing in the vats is due to the fact that the eighth post bears directly on the wall plate of the centre pier in the space between the two rows of vats. Where the posts rest on the transverse joists direct they are checked into them, but those standing in the vats bear on hard wood blocks 5 in. thick.

The posts, which are supported in places by 8 in. by 14 in. and 9 by 6 in. braces, carry 14 by 14 in. timbers running the length of the plant, these being tied together between every pair of vats by a 12 by 14 in. transverse beam. These 14 by 14 in. longitudinal timbers carry the upper vats on a series of transverse joists spaced similarly to those supporting the lower vats but of different dimensions—viz., 12 by 12 in., 8 by 12 in., and 6 by 12 in. The upper vats do not rest directly on these joists but on a series of 9 by 3 in. deals crossing them and placed with 15 in. spaces, except the six deals running alongside the discharge doors which are 9 by 6 in. placed with 2 ft. spaces.

The vertical distance between the rims of the lower vats and the staves of the upper ones is 5 ft. 6 in. and platforms permit of free access to all parts of the plant.

The launder into which the slime and water pass from the intermediate vats convey it to a spitzkasten where any sand in suspension settles and is returned to the sump of the tailings pump. The spitzkasten, which is of the usual pyramidal shape, is 6 ft. deep by 6 ft. by 6 ft. at the top, and 12 in. square at the bottom which has a 2 in. outlet. The slimes pass from this spitzkasten to the dam where they are being conserved pending the erection of a plant for their treatment. The tracks under the leaching vats for the trucks which convey the residues to the dump are of 16 lb. rails with a gauge of 18 in. The track has a fall of 12 in. from one end of the vats to the other to facilitate the handling of the trucks. These latter are side tipping, of 20 cubic feet capacity, and are run by a $\frac{5}{8}$ in. endless wire rope at the rate of $1\frac{1}{2}$ mile per hour, the length of haul being about 900 ft. and the engine a Tangye horizontal non-condensing with a 10 in. cylinder and 22 in. stroke. The trucks enter at one end of the series of vats and emerge at the other, there being an iron floor at each end of the plant.

The extractor house of this plant is worthy of special notice. The building is 100 ft. long by 51 ft. wide and 24 ft. high to the ridge, and besides the precipitating plant, &c., contains two offices and a storeroom. An independent $1\frac{1}{2}$ in. pipe runs from the bottom of each of the twenty-eight vats to this house where they terminate in a horizontal row of stopcocks grouped $7\frac{1}{2}$ in. from centre to centre about 5 ft. 6 in. from the floor and overhanging a rectangular wood tank divided by vertical partitions so as to form three separate coir filter boxes, one each for the weak, the medium, and the strong solutions, the size of each box being 7 ft. 6 in. by 7 ft. 6 in. by 4 ft. 6 in. deep. The coir compartment of each of these boxes is 7 ft. 6 in. by 5 ft. 8 in., and has a false bottom 9 in. deep covered with one thickness of cocoanut matting, whilst a light wood framework rests on the coir. The cyanide solutions gravitate from the vats to these coir boxes through an intermediate three-compartment launder running across the latter, each compartment of this launder being perforated over one of the filter boxes. Short lengths of hose-pipe convey the solutions from the stopcocks to any of the launders.

From each of these coir boxes proceeds a 3 in. pipe, these connecting with a 3 in. main which runs at the head of the series of nine extractor boxes and is so provided with stopcocks that different strengths of solution may be run into different extractor boxes at the same time.

The latter, which are arranged side by side, are each 23 ft. long by 3 ft. wide and 2 ft. 6 in. deep internally, being flat-bottomed. They each contain seven zinc compartments and two settling compartments, one at the head and the other at the foot, the latter also being used for dissolving the cyanide. The boxes are constructed of 1½ in. clear pine boards with 3 by 2½ in. posts and a top rail of 6 by 3 in., the divisions being of ⅜ in. board.

The trays for carrying the zinc thread have ⅛ in. screens, and are 6 in. deep. On two of their sides is an iron plate with two holes in it to take the ends of the hooks by means of which the trays are raised. Projecting handles, which are an inconvenience at the clean-up, are thus done away with. At the bottom of each of the box partitions is a plugged hole through which, at the clean-up, the gold slime is washed to the end compartment where it is collected.

Each box, the top four compartments of which are provided with a cover, rests on two longitudinal deals bearing on the sloping cement floor of the extractor house the fall of the boxes being 3½ in. in their length. The solutions leave each box by a 3 in. pipe connecting with a common 3 in. main, this latter having three branches leading to three 3 in. pipes running to the sumps, all these last pipes having a connection with each of the sumps. The cement floor is guttered around the extractor boxes, the channels conveying all leakage to a small well from which a Cameron pump transfers it to a settling vat outside the building.

The tailings are, in the ordinary course, to be leached by gravity but provision is made for expediting by vacuum when necessary, the leaching pipes of all the vats being connected with a vertical vacuum cylinder 7 ft. high by 4 ft. diameter. The pump which exhausts this cylinder is a Howard-Farrar geared double-action horizontal with a 6 in. ram, this delivering to the three coir filter boxes. The draining of the water from the intermediate vats can also be assisted by this vacuum.

The solution pumps, which are near the above, are two 4 in. centrifugals of Gwynne's make. One of these is kept as a standby, a single pump being found sufficient to transfer all solutions from the sumps to the intermediate and leaching vats. The pumps have a common 4 in. suction main which has a 4 in. branch with stopcock and foot valve in each sump. The method of distributing the solutions to the vats is novel. The 4 in. delivery pipe from the pump discharges into any one of four circular iron pans, 3 ft. diameter and 1 ft. 6 in. deep, three of which are divided by radial partitions into eight compartments and the

Glencairn Main Reef Gold Mining Company, Limited.

View of Extractor House, showing Filters, Extractor Boxes, and Pumps.

other into four compartments, a pipe 4 in. diameter proceeding from the bottom of each of the compartments to one of the vats, each distributing pan being placed just above the level of the intermediate vats and near the particular group of vats it supplies. The pump delivery pipe passes up the centre of the pan and has a swivel discharge into it. The water supply for the vats, which comes direct from a reservoir, also passes through these pans, the main being 4 in. diameter. This method of distributing the water and solutions does away with a large number of stopcocks.

The solution and vacuum pumps together with a plunger pump, zinc lathe, slag crusher, and grindstone, are driven from a 3 in. shaft actuated by a motor designed to give 200 volts 170 amperes at 830 revolutions per minute. The plunger pump is a three-throw geared one with 10 in. plungers and 14 in. stroke, its work being to lift water from the slimes dam to the reservoir. The zinc lathe is of the double hand type, taking 10 in. discs, and the slag crusher is a small Ball mill.

There are three iron sump vats—for the weak, medium, and strong solutions—36 ft. diameter and 8 ft. 9 in. deep, placed at such a level that the solutions from the extractor boxes gravitate to them. Their sides, which are of $\frac{3}{16}$ in. iron, are two plates deep with an angle iron $2\frac{1}{2}$ by $2\frac{1}{2}$ in. by $\frac{1}{4}$ in. at the top and bottom. The bottom plates, which have parallel joints, rest directly on the bottom of an excavation. The strong solution sump is provided with two rotating steam agitators each consisting of two arms, of perforated pipe, free to revolve in a horizontal plane on a vertical steam supply pipe; the reaction due to the efflux of the steam through the holes in the arms causes the latter to rotate.

The arrangements for the clean-up are very complete, the apparatus being situated in one of the corners of the extractor house. There is a 2 in. Cameron pump to deliver the supernatent solution from the extractor boxes to a settling vat 13 ft. 6 in. diameter by 7 ft. deep outside the building and above the level of the boxes, this vat having a number of holes down its side through which the solution, after settling, is decanted back to the boxes. The corroded zinc and the gold slime is transferred from the boxes to a screen resting on the coarse slime vat (5 ft. 6 in. diameter by 2 ft. 6 in. deep), the screen being of $\frac{3}{16}$ in. mesh. Alongside this vat is another of similar dimensions, the filter press vat, to which all the gold slime is ultimately transferred and which has a small pipe well 6 in. diameter in its bottom connected by a $1\frac{1}{4}$ in. suction pipe with the filter press. The latter is of Johnson's patent make with plates $18\frac{1}{2}$ in. square and carries the steam pump on its frame, the slimes plunger being $2\frac{1}{2}$ in. diameter. The press, together with the two clean-up vats described above and the coir solution filter vats, stands in a cement pit with an inclined bottom about 12 in. deep, the slope being to a well

E

10 ft. 7 in. by 10 ft. 7 in. with a depth of 2 ft. 6 in. to 2 ft. 9 in. the bottom of this being also inclined. In this cement well stands the vat for the decanted solutions of the acid vat, which latter is supported immediately above it, the first vat being 10 ft. diameter and 4 ft. 6 in. deep with holes down its side for decanting, and the latter vat 7 ft. diameter by 6 ft. deep, this having a similar provision for decanting the spent acid and washes into the vat below. The acid vat, which has a 2½ in. pipe to the slimes filter press vat, is provided with a hinged cover and has a flue to convey the acid fumes out of the building; it is also fitted with a wooden agitator to be worked by hand. The zinc and acid are not introduced into the vat direct but are passed into it through an intermediate box with a sliding bottom.

The press filtrate runs to a small sump 4 ft. by 6 ft. by 3 ft. deep.

The roasting and smelting appliances are in the same building as the general offices. There is one gold slime roasting furnace with a fire grate 3 ft. by 3 ft. under a ¾ in. wrought iron plate, 6 ft. 6 in. by 3 ft., which forms the hearth, this plate being rivetted to three 45 lb. stiffening rails running longitudinally. The roof of the roasting chamber is arched, being 10 in. high at the sides and 21 in. at the centre. Alongside this is the slime smelting furnace which is of the reverberatory type, having a grate area 3 ft. by 3 ft. and a crucible hearth 5 ft. 9 in. by 3 ft. at a level 9 in. higher than that of the grate. The roof of the crucible hearth is 1 ft. 5 in. high at the sides and 1 ft. 9 in. at the centre and the door to it is 2 ft. by 16 in.

For the melting of both cyanide and mill gold and running into ingots there are two circular furnaces 17 in. diameter by 2 ft. 1 in. deep.

The amalgam retort, which is in the same room as the above furnaces, is 9½ in. diameter by 3 ft. 6 in. long.

In the assay room, adjoining the smelting room, are two fusion furnaces 16 by 16 in. by 2 ft. deep at the front, the tops having a slope, also a Morgan E muffle.

This completes the description of this model outcrop installation, the design and erection of which are creditable to both Mr. Blane and Mr. Roberts the engineer. The capacity of the mill is expected to be about 850 tons per 24 hours and of this approximately 70 per cent. will be caught as sands and concentrates for treatment by cyanide.

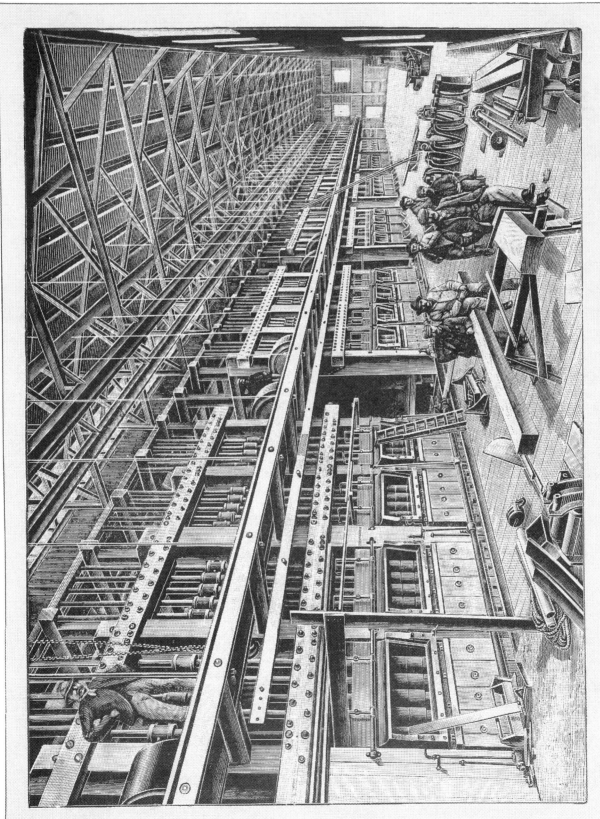

Simmer and Jack Proprietary Mines.—280 Stamp Mill.

THE SIMMER AND JACK PROPRIETARY MINES, LIMITED.

TO Mr. J. H. Hammond we owe the conception and the bringing into existence of the great installation on this property—a property embracing 1,206 claims, of which 649 will be worked by the company, the remaining 597 claims having been merged into sub-companies and being held on a scrip basis.

Though possessing large deep level interests the company is an outcrop venture and its claims extend for nearly a mile along the reef.

The mill and cyanide works, &c., the largest of their kind in the world, have been engineered by Messrs. Connor and Webb, whilst the general management is in the hands of Major Sapte and Mr. F. M. Watson. When its 280 stamps are dropping the company will have about 450 white employees and 3,500 to 4,000 natives and the total monthly wages will amount to about £20,000.

The property is being worked from eight shafts one of which, the east incline, follows the reef from the surface; the others cut the reef vertically at depth and then go down on the underlay. At present ore is being hauled at two of these shafts, the north vertical shaft and the east incline.

The north vertical shaft has two skip ways, a ladder way, and a pump compartment. The skips, which are of 35 cubic feet capacity and automatic tipping, discharge on to a grizzly from which the coarse rock drops to a washing floor and then goes to the sorting belt which conveys it to the crusher, this being a No. 5 Gates. The sorting belt is 46 ft. between centres of end pulleys and 2 ft. 5⅛ in. wide. The bin into which the crusher passes the ore discharges into great trucks of 20 tons capacity which run beneath it, these carrying the ore to the mill.

The crushing and sorting arrangements at the east incline shaft are a duplicate of the above and the other shafts will be laid out on similar lines.

The 20 ton trucks here mentioned have their two ends sloping inwards to their bottom discharge door which is the full width of the truck.

They run on a track of 3 ft. 9 in. gauge, a steam locomotive hauling them from the shafts to the top of the mill ore bin, the grade into the mill being 3 per cent. ; the locomotive, which has six wheels coupled, has two 15 by 20 in. cylinders.

THE MILL.

The mill building is 100 ft. wide by 270 ft. long and is 33 ft. 8 in. from the floor to the lowest tie beam of the roof. The 280 stamps, of 1,250 lbs. weight, are arranged in two rows, back to back, in batteries of twenty, with a space of 11 ft. between the adjoining king posts of each twenty.

The line sills, mud sills, mortar piles, and the lower timbers of the ore bin are Karri wood, and the upper part of the bin and the battery posts are of Oregon pine.

The ore bin is flat-bottomed and is 31 ft. wide by 16 ft. deep. Its mud sills, of which there are seven, are of 14 by 14 in. section, three of them resting on a 6 ft. masonry pier in the centre, the other four being carried by side piers, 9 ft. wide, two on each. All are securely anchored to the masonry which is set in cement and rests on concrete footings. The transverse sills crossing these mud sills, and which form the line sills of the battery, are 16 by 12 in. timbers except those carrying the centre king post of each battery of 20 stamps which are 16 by 20 in. These transverse sills, between which there are 5 ft. spaces, carry the 12 by 14 in. bin foundation posts, three on each, and the four 12 by 12 in. struts, two in each frame. On these posts rest the three 12 by 14 in. longitudinal sills which carry the 14 by 12 in. bottom joists of the bin, the prolongations of these joists forming the knee beams of the battery framing. Each of these joists in turn has resting on it two 12 by 12 in. bin side posts and one 14 by 12 in. centre post, also four 12 by 9 in. struts, these posts having an upper 14 by 12 in. transverse tie beam on which, over each side post, rests an 8 by 8 in. longitudinal tie, whilst the four 12 by 12 in. runners carrying the rails also lie on it. The bolts used throughout the framing are of 1 in. and 1¼ in. diameter. The sides and bottom of the bin are of 12 by 3 in. deals these being lined on the sides with 1 in. boards and on the bottom with 2 in. boards.

From centre to centre of the mortar blocks of the two lines of stamps is 47 ft. The blocks are built up of eight 14 by 15 in. piles 15 ft. long bolted together by eighteen 1 in. bolts and having three hard wood transverse keys. They rest on concrete and are between two masonry walls of which the front one is 3 ft. thick, the back being the 9 ft. masonry pier carrying the side mud sills of the ore bin. The depth of the pit thus formed is 8 ft. 7½ in. and it is filled in, all around the mortar blocks, with concrete. The mortar blocks project 4 ft. above the line sills and have their upper end supported by a 10 by 10 in. timber at both back and front, these resting on the line sills and being bolted to the battery post. The line sills—the continuation of the lower transverse sills of the ore bin—are supported in front and behind the mortar blocks on a 16 by 14 in. mud sill resting on the masonry walls. To every 20 stamps there are four 2 ft. by 12 in. battery posts and a centre one of 2 ft. by 20 in. section, the height being 21 ft. 8 in. They are each braced to the knee post—the bottom joist of the bin—and by a 9 by 12 in. timber to the ore bin posts, whilst to the line sills they are secured by two vertical bolts, one at the back and one at the front of each post, passing through the ends of a bar running horizontally through the latter.

Laterally the battery posts are supported by the two guide beams of 14 by 12 in. section placed 6 ft. 6 in. apart, the bottom of the lower one being 10 ft. 4 in. above the line sills.

The knee beam, which is 12 in. below the centre of the cam shaft, carries the cam floor this having a width of 6 ft., whilst in front of the posts is another floor 12 in. below this. The feeder floor is flush with the top of the mortar blocks. The guides consist of a light cast steel frame which has five recesses into each of which fit maple guide blocks, these blocks being each held in position by a front casting wedged up by two oak keys, the latter being prevented from working out by a bow-spring the ends of which are forced into them by a bolt fixed to the front casting.

The mortars are of the Alaska-Treadwell type. They are 58 in. long, 18 in. broad (at the top), and 56 in. high, and have a feed inlet 54¼ in. long by 3½ in. widening to 4 in. The thickness of the mortar bottom is 9 in. and there are eight holding-down bolts, 1½ in. diameter and 41 in. long, with 4 in. washers at the lower end. The height of the discharge aperture is 20½ in. and from the bottom of the mortar (inside) to the inner and lower lip of this is 5½ in.

There are cast steel liners of special design—recessed—on the back and ends, whilst the front has a plain steel liner and the feed inlet an iron liner. The breadth of the mortars inside the linings is 10 in. at the bottom and 13¼ in. at the level of the top of the dies. The screen, which is of 2 by 2 in. framing with an iron facing for the keys and presents one aperture 51¾ in. long by 8 in. deep, rests on chuck blocks of which there are three sizes all carrying ¼ in. amalgamated copper plates. The screening used will be about 600 mesh. From the centre of the mortar to the outer edge of its discharge lip is 16 in. and to the latter is bolted a cast-iron apron 14 in. long, with a slope of 2⅛ in., which has two slots for the pulp 24¼ in. by ¾ in. The mortars are fed by suspended challenge feeders.

The dies of forged steel weigh 134 lbs. and have a cylindrical body on a square base, the former being 9 in. diameter by 6 in. high, and the latter 9⅞ in. square by 1 in. thick. They are placed with their centres 10 in. apart.

The stems, which are of faggoted iron, are 16 ft. long and 3½ in. diameter, a 6 in. taper at the end reducing them to 3¼ in. They weigh 571 lbs.

The cast steel tappets weigh 119 lbs. and are 13 in. high by 9¼ in. diameter with a 2½ in. face. The gibs are wedged by three keys. The heads, also of cast steel, scale 274 lbs. and are 18 in. high and 9 in. diameter, whilst the shoes, which are of forged steel, have a butt 9 in. high and 9 in. diameter, with a shank 4¼ in. to 3¼ in. diameter and 5¼ in. high, the weight being 160 lbs.

The stamps, which are arranged as before described, in batteries of twenty, have a cam shaft for each ten, these shafts, of faggoted iron, being 6½ in. in diameter and 14 ft. 3 in. long carrying a wood pulley 6 ft. diameter with a 17 in. face this being covered with ¼ in. iron. The cams, of cast steel, are 30 in. over the toes and have a 2½ in. face and a boss 11½ in. diameter by 5 in. long. They are secured on the shaft by two ordinary keys, the distance between the centre of the stems and the side of the cams being 2 in., whilst from the centre of the stems to the centre of the cam shaft is 5¾ in. The order of drop is 1, 10, 5, 6, 2, 9, 4, 7, 3, 8.

The cam shaft bearings are of the open top type, the centre one being 20½ in. long and the others 12½ in. long.

The Jack shafts are 3 in. diameter and a crawl is arranged to travel from the mill store to any part of the battery.

Resting on the lower bar of the screen frame is a ₁¹₆ in. amalgamated copper plate, 15 in. long and the width of the screen, set with the same slope as the cast-iron apron. This plate delivers the pulp to a similar one sloping down towards the mortar and having its forward lip curved up to prevent splash. From the lower edge of this the pulp drops to the iron apron

(43)

Simmer and Jack Proprietary Mines Cyanide Works.

View showing General Arrangement.

F

Simmer and Jack Proprietary Mines Cyanide Works.

Showing Leaching Vats, Extractor House, and Pipe System, with Intermediate Vats in the Distance.

from which it passes to the amalgamated tables 16 ft. long and 4 ft. 9½ in. wide. These are in two sections with a step of 3 in. between them, the upper table being 10 ft. by 4 ft. 9½ in., and the lower one 6 ft. by 4 ft. 9½ in., an amalgamated splash plate 13 in. deep being placed at the step. The plates, which are of ⅛ in. copper, are set with a fall of 1¾ in. to the foot and at the bottom of each table is a cross launder with a 2 in. pipe leading the pulp to a mercury trap from which it passes by a launder with a 4 per cent. grade to three elevating wheels situated just outside the mill. The clean-up appliances include five pans, 36½ in. diameter by 13½ in. deep, and amalgamated plates.

The stamps, which will make about 100 drops of 9 in. per minute, are driven from two lay shafts varying from 8 in. to 4 in. diameter, these having their bearings on the line sills immediately behind the mortar blocks. Belt tighteners are used to throw the stamps into action.

The mill engine is a 900 I.H.P. Yates and Thom compound, tandem, surface-condensing horizontal, with Corliss valve gear, the cylinders being 24 in. and 44 in. diameter with 54 in. stroke and the driving being by ropes.

The elevating wheels outside the mill are 38 ft. diameter and lift the pulp 28 ft. to the launder running to the cyanide works. They make 4·5 revolutions per minute and each has 103 buckets 12 in. deep and 16 in. wide tangent to a circle 27 ft. 3 in. diameter, the pitch being 13·9 in. Each wheel has sixteen arms of 5 by 9 in. timber and these are each stayed by two 1½ in. bolts one on each side. The wheel bearings, which are 10 ft. from centre to centre, rest on 14 by 16 in. sills carried by masonry pillars. The bottom of the buckets—the outer circumference of the wheel—consists of ⅛ in. iron plate. Fixed to the arms of the wheel is the run, 25 ft. diameter and 18 in. face, for the belt which drives the wheel.

THE CYANIDE WORKS.

The difference in level between the mill and these works is so small that it necessitated the elevating of the pulp at the mill by the above wheels. The launder from the latter carries it to three other wheels of similar design but of larger dimensions at the cyanide works, the diameter being 42 ft., the number of buckets 132, the belt drive 30 ft. diameter with a 17 in. face, and the shaft 10 in. diameter, whilst the number of revolutions is 3½ per minute, the driving belts being of Dick's make and 15½ in. wide. The launders into which these wheels discharge run to a distributing box from which the pulp passes to two parallel series of spitzluten, three boxes in each series, between which run four launders which form a by-pass in the event of the spitzluten going wrong.

The spitzluten are each 3 ft. 1 in. deep from the inlet lip to the bottom by 4 ft. 1 in. long and 1 ft. 10 in. wide, with a base $4\frac{1}{2}$ in. square containing a $2\frac{1}{2}$ in. outlet the water pipe to this being $2\frac{1}{2}$ in. diameter and the concentrates discharge nozzle tapering to $1\frac{1}{4}$ in. diameter.

The long sides of the boxes are perpendicular to 1 ft. 7 in. below the inlet lip and the total depth of the sides is 4 ft. 1 in. the boxes being constructed of 3 in. deals with baffle boards 2 in. thick. The concentrates pass from the bottom of these boxes by launders having 4 per cent. and 6 per cent. grade to the intermediate concentrate vats, ten in number, whilst the sand and slime pass on by other launders to the sand intermediate vats, of which there are twenty, where they deliver upon rotary distributors.

These thirty intermediate vats, ten for concentrates and twenty for sand, are arranged in five parallel rows of six vats each and have their centres 32 ft. apart the short way and 28 ft. the long way of the plant, each vat being stepped down 6 in. in this latter direction. They are of wood and are 24 ft. diameter and 11 ft. 3 in. deep—internal dimensions—with 3 in. staves and bottoms, and each has eight hoops $1\frac{1}{8}$ in. diameter with $1\frac{1}{4}$ in. ends. Each bottom has the ordinary filter and carries six discharge doors, these being arranged in two rows of three each, over the two discharge ways which run beneath each row of vats. Around the rim of each vat is an outer annular launder 6 in. wide and 12 in. deep having its bottom and outer side of 3 in. deals and being encircled by two hoops, the upper one $\frac{3}{8}$ in. diameter, and the lower one $1\frac{1}{8}$ in. diameter; into this launder the slime and water overflow, baffle boards being fixed to the rim of the vat on the inner side to prevent a direct flow, the boards projecting 2 in. below the rim of the vat. From these annular launders the slime and water are conducted to a series of six spitzkasten where any sand in suspension is deposited and is returned to the wheels; the slimes at present flow from the boxes to a dam but they will shortly be deflected to the slimes plant now in course of erection. The six spitzkasten are each 13 ft. long by 4 ft. wide at the top end and 5 ft. wide at the bottom end with a depth of 4 ft. and 5 ft. at the inlet and outlet end respectively, the boxes being set with their tops horizontal. The bottom is divided into five pyramidal compartments, the divisions being about 2 ft. high, and from the bottom of each of these compartments runs a $\frac{3}{4}$ in. outlet pipe which delivers the sand by a hose into a sloping launder 2 ft. above the bottom of the spitzkasten, this carrying the sands to the tailings wheel.

The bottom of each intermediate vat rests on 6 in. by 12 in. joists placed with 12 in. spaces except those beside the discharge doors which are 24 in. apart; these joists bear on

(49)

Simmer and Jack Proprietary Mines.

Cyanide Works, Tailings Wheels, Spitzluten, and Intermediate Vats.

12 by 6 in. timbers lying flat on the three masonry piers which carry each vat. The side piers are 21 ft. long by 4 ft. 6 in. wide at the centre, this tapering to 2 ft. at each end, and the centre pier is 25 ft. long by 4 ft. wide at the top and 4 ft. 6 in. at the bottom. These piers form two passages 5 ft. 9 in. wide at the top and about 6 ft. high which have their centres 11 ft. apart, the bottom having a slight down grade in the direction of the leaching vats. The discharge trucks are hauled on the single rope system up inclined trestles to the platforms over the leaching vats, the grade of the former being 20 per cent. whilst the latter also has a grade of 3 per cent. to permit the trucks to return by gravity.

There are two tracks over each row of leaching vats with an electric hauling engine to each, these being at the platform level.

The leaching vats, thirty in number, ten for concentrates and twenty for sands, are arranged similarly to the intermediate vats being in five rows of six vats each, each row having the same centre line as a row of intermediate vats and the vat centres being 32 ft. apart the long way of the plant. These vats, which are also of wood, are each 30 ft. diameter and 10 ft. deep and are of 3 in. timber with nine $1\frac{1}{8}$ in. diameter binding hoops, each vat bottom having four discharging doors.

They rest on joists and masonry walls similar to those of the intermediate vats except that instead of one centre pier there are two, each 2 ft. thick and 3 ft. apart. The vats are each stepped down 6 in. in the direction of both the long and short axes of the plant, those at the far end of the plant being the highest. The twelve tracks for the discharge trucks, two under each row of five vats, join on to one main line running with a 6 per cent. grade to the dump, the return line having a grade varying from 6 per cent. to $1\frac{1}{2}$ per cent., the hauling being by endless rope. The vats have the usual filter bottom and are provided with shields on their rims to prevent any sand being tipped outside them.

Each intermediate and leaching vat has an independent $1\frac{1}{2}$ in. leaching pipe which runs to a testing house and a 4 in. pipe supplies water to the vats whilst a similar pipe carries the solutions to them.

In the testing house the leaching pipes discharge into a distributing box from which the solutions gravitate by three launders to the three lower sumps. This testing house contains a three-throw electric pump with 4 in. plungers for passing solutions direct to the extractor boxes when desired. The three lower sumps are 24 ft. diameter and 7 ft. 3 in. deep internally. They are of wood and are placed in line standing on the cemented bottom of an excavation with their centres 27 ft. apart. Alongside them is a vat into which the prussian blue from the extractor boxes is run its dimensions being 16 ft. diameter by 5 ft. 6 in. deep.

G

From these lower sumps the solutions are lifted by three 3 in. centrifugal pumps and one 2 in. centrifugal to the extractor boxes which they enter by a 2 in. branch from the 3 in. main, each pump supplying a group of the boxes.

Another 3 in. centrifugal pump alongside the above returns any leakage about the extractor house and sumps to the upper sumps.

The pumps for supplying the intermediate and leaching vats with solution and water are three 4 in. centrifugals with a 4 in. suction pipe to each of the upper sumps and two 4 in. delivery mains, one for water and one for solution. All these pumps are located in an annexe (43 ft. by 30 ft.) of the extractor house and are driven by electricity.

The extractor house is 170 ft. long by 42 ft. wide and has a cement floor. It contains sixteen Siemens-Halske electric precipitation boxes the current for which is generated by a dynamo near the pumps. The boxes, each 30 ft. long by 4 ft. 9 in. wide by 3 ft. deep—internal dimensions—are placed with their centres 10 ft. apart and are divided into ten main compartments 2 ft. 7 in. by 4 ft. 9 in. by 3 ft. deep, and are like zinc extractor boxes in general design the solution passing upwards and downwards alternately. In the ten compartments are the vertical iron anodes, 2 ft. $6\frac{3}{4}$ in. by 2 ft. 3 in. by $\frac{3}{16}$ in. thick, placed with their centres 4 in. apart the plates running the long way of the boxes ; the cathodes, which consist of lead shavings supported by light frames, lie between them. The current is conveyed to and from the electrodes by two copper strips, 2 in. by $\frac{1}{8}$ in., one on each side of the box, each of these conductors connecting with alternate copper ribbons, 1 in. by $\frac{1}{8}$ in., which cross the box at each partition and pass through mercury wells 1 in. diameter by $1\frac{1}{4}$ in. deep into which the wire connections of the electrodes dip.

At the bottom of the side of each of the ten compartments is an exit for the prussian blue which flows into a launder running alongside each box, these conveying it to the prussian blue vat.

The extractor boxes are built of 3 in. timber with $1\frac{1}{4}$ in. partitions, the sides being supported by $4\frac{1}{2}$ in. by 3 in. posts and the bottom by $4\frac{1}{2}$ in. by 3 in. deals which rest on three 4 in. by 6 in. bearers carried by the floor. The solution leaves them by a 3 in. outlet and flows into a small open vessel to the bottom of which the 4 in. mains running to the upper sumps are connected. Metallic connection between the boxes and the sumps is avoided as it was found that with it there was a loss of electricity.

The three upper sump vats are 24 ft. diameter and 7 ft. 3 in. deep and are of wood. To these vats, which are placed just outside the extractor house with their centres 27 ft. apart,

Simmer and Jack Proprietary Mines Cyanide Works.

Tailings Wheels, and Intermediate Vats with Sloping Tram Track to Leaching Vats.

the solutions from the extractor boxes gravitate, and after they are made up to the required strengths they are returned to the intermediate and leaching vats.

Adjoining the extractor house is a shed containing two Liddell's automatic turning machines for making the lead shavings which form the cathodes.

The furnace house of this plant contains two cupellation furnaces for the lead cathodes, one lead melting furnace, two amalgam retorts, and four fusion furnaces.

Apart from the size and great capacity of this Simmer and Jack Proprietary Mines installation it has many features which attract attention and it stands a fitting testimonial to the enterprise of the Rand.

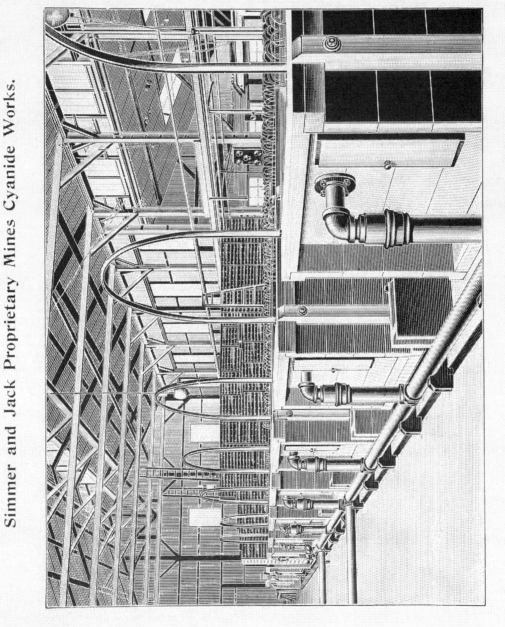

Simmer and Jack Proprietary Mines Cyanide Works.

Interior of Extractor House, showing the Electrical Precipitation Boxes.

JUMPERS DEEP GOLD MINING COMPANY, LIMITED.

THE Rand is yet new to deep level mining. The deep properties now at work are, with one or two exceptions, still in the developing stage, and, therefore, with such limited experience of the mining and metallurgy pertaining to the deeper zones of the reefs, there may remain much for us to discover in this connection. We still, however, have to learn that the conditions obtaining with the reefs at depth are such as to necessitate surface installations essentially different from those of the outcrop companies, a contingency which, so far as the first row of deeps is concerned, is very remote, for no hard-and-fast line can well exist between these and the outcrop companies.

The Jumpers Deep Company, Mr. Strangman Hancock, manager, the installation of which is here described, is one of the holdings of the Rand Mines (Limited), whose properties are under the general management of Mr. G. E. Webber with Mr. L. J. Seymour as consulting mechanical engineer, and the plant when completed will be one of the finest on the field.

The 325 claims are to be worked from two rectangular shafts, 21 ft. by 6 ft. in the clear, which are vertical to the intersection of the reefs and thence inclined, the depth of the vertical section of No. 1 shaft being 1,006 ft. and that of No. 2 shaft 1,306 ft., whilst the angle of the underlay is about $28\frac{1}{2}°$.

Each headgear is 60 ft. from the masonry foundations to the top sills, the four legs, of 14 by 12 in. section, resting on 12 by 14 in. sills which bear on masonry foundations to which they are anchored by $1\frac{1}{8}$ in. bolts. The braces are 10 by 10 in. and the framing timbers

H

12 by 12 in. with tie bolts 1 in. to 1¼ in. diameter, whilst the back stay legs are 14 by 12 in. with 10 by 10 in. braces and 8 by 10 in. framing. The shafts have each three winding compartments, 4 ft. 6 in. by 6 ft., a pump compartment 4 ft. 5 in. by 6 ft., and a ladder way 16 in. by 6 ft., all clear dimensions, whilst the winding sheaves are 12 ft. diameter and have their bearings on 12 by 18 in. timbers.

The skips, of $\frac{5}{16}$ in. steel with liners, are of the usual automatic tipping type and have a capacity of 3 tons. They discharge on to the cover of the waste bin, sliding thence to the grizzly, 21 ft. by 12 ft., set with an angle 40° and constructed of manganese steel bars 3½ in. deep by 1 in. thick at the top and ½ in. at the bottom, the spaces at the top being 1⅜ in. From these bars the coarse rock passes into the coarse bin, the fines passing the bars going to a separate bin. There are no sorting or crushing arrangements in the headgears, these operations being performed in an independent crushing tower.

There are two winding engines at No. 1 shaft: one by the Edward P. Allis Company, of Milwaukee—a single cylinder (18 by 48 in.) horizontal with Corliss valve gear and 8 ft. drum, capable of hauling a load of 3 tons from a depth of 2,000 ft. in 70 seconds; the other by Cochrane, of the direct-acting, cross-coupled type, with Corliss valve gear, the cylinders being each 20 in. diameter by 48 in. stroke. This engine is designed to lift a load of 5 tons from a depth of 2,000 ft. in one minute.

At No. 2 shaft one engine is a Yates and Thom, 20 by 60 in., single cylinder, with two drums 8 ft. diameter, and the other is a Fraser and Chalmers' single-drum direct-acting, also with a 20 by 60 in. cylinder.

The ore, both coarse and fines—these being kept separate—is hauled from the headgears by endless rope to the crushing and sorting tower situated about midway between the two shafts. The top floor of this to which the trucks are carried is 36 ft. above ground level and has an area of 45 ft. by 36 ft. Here the coarse is tipped into a bin from which it drops to a circular washing and sorting table,* whilst the fines are passed through a revolving screen the rejections of which go to a Blake's crusher. The sorting table is an annular ring, 25 ft. outside diameter and 4 ft. wide, of ⅜ in. iron plate and carries beneath it flanged wheels which run on a circular rail, the table being driven by rack and pinion the latter having a friction clutch. Under the table is the bin into which the worthless stone is thrown and falls to a double tram track leading to the waste dump, whilst the ore is carried round to three E size Comet crushers (24 by 12 in.) placed over the supply bin, 52 ft. 6 in. long by 16 ft. wide by 8 ft. deep to the apex, the two sides meeting and forming an angle of 90°, this having sixteen discharge

* This arrangement has been remodelled and two tables erected.

Jumpers Deep Gold Mining Company, Limited.

View of one of the Sorting Tables showing the Screening and Washing Cylinder.

doors opening on to the two tram tracks beneath it. The ore is conveyed from here to the mill in box trucks of 40 cubic feet capacity, the haulage being on the single rope system.

THE MILL.

The building is 203 ft. long by 75 ft. wide and 34 ft. 9 in. high from the line sills to the tie beams of the roof trusses. There are 200 stamps,* of 1,050 lbs., by Fraser and Chalmers (Limited), arranged in two rows of 100 each back to back, with intermediate ore bin, from centre to centre of the two lines of mortar blocks being 36 ft. The stamps, which are arranged in batteries of ten, have their mortar blocks resting on 12 in. of concrete between two masonry walls 8 ft. 9 in. high running the length of the mill. These walls, which are of large roughly squared stone set in cement mortar, are 3 ft. 6 in. thick at the top and 4 ft. 6 in. at the bottom, and form a pit 3 ft. 6 in. to 4 ft. 6 in. wide for the mortar blocks which are packed in them by tightly rammed sand, each block having a ½ in. layer of a mixture of tar and sand around it. The tops of the masonry walls are 2 ft. 9 in. below the centre of the mill shaft, and the two inside walls have a step 2 ft. deep and 4 ft. wide to take the two ore bin mud sills. The mortar blocks, line sills, mud sills, and battery posts of this mill are of pitch pine and the rest of the mill of Oregon pine.

The mortar blocks, 5 ft. by 2 ft. 6 in. by 14 ft. 6 in. deep, each consists of four end piles: two 15½ in. by 18 in., and two 15½ in. by 12 in., and four intermediate: two 14½ in. by 18 in. and two 14½ in. by 12 in. These piles are bound together by six 1¼ in. bolts and twelve 1 in. bolts and have nine 4 by 4 in. hardwood keys which are arranged in three rows. At their upper end, which is 4 ft. 3 in. above the centre of the lay shaft, they are supported by 9 by 9 in. cross timbers bolted with 1 in. bolts to the battery posts, there being two of these timbers at the front of the block and two at the back, the upper one being 1 in. from the top of the block whilst the other is 2 ft. 2 in. below this one and rests on the line sills. The latter each consists of two timbers bolted together, an upper one 16 in. deep and a lower one 12 in. deep, the length of both being 12 ft. 4 in.—6 ft. 2 in. in front of the block centre line and 6 ft. 2 in. behind. The breadth of those carrying the pulley king post of each battery of ten stamps is 14 in. whilst that of the others is 12 in., and besides the five 1½ in. bolts with which the two timbers of each sill are bolted together they have three hardwood keys 7 in. by 2 in. Two of the five 1½ in. bolts in each line sill pass through the battery mud sills, two in number, one bolt through each; these mud sills—one 3 ft. 3 in. in front and one 3 ft. 3 in. behind the centre of the mortar blocks—being 18 in. by 14 in. timbers

---

* One hundred of these stamps have been erected.

anchored by 1¼ in. bolts, 6 ft. long and 6 ft. 0½ in. and 6 ft. 2 in. apart, to the two masonry walls. The centres of the adjoining line sills of each battery of ten stamps are 6 ft. 2 in. apart; the others have a 6 ft. 0½ in. pitch.

The battery posts are 21 ft. 3½ in. high, the pulley post being of 24 in. by 14 in section and the others 24 by 12 in., there being one cam shaft and pulley to each ten stamps. The distance between the posts is 5 ft. and each is secured to the line sills by two 1½ in. bolts running diagonally. There are two guide beams, one 16 by 10 in. placed 17 in. from the top of the posts and the other 18 in. by 10 in. fixed 9 ft. 9 in. from the top, and the adjoining posts of each ten stamps are braced together at the top by a 10 by 6 in. timber. Each battery post has a 12 by 10 in. knee beam 1 ft. 7 in. below the centre of the cam shaft, this being the continuation of the ore bin transverse sill, and at the top there is a lighter brace 10 by 6 in. to the bin post.

The mortars are 4 ft. 10½ in. long by 19 in. wide at the top by 4 ft. 10½ in. deep, all external dimensions, and each weighs about 7,250 lbs. They are secured to the mortar blocks by eight 1½ in. bolts, 5 ft. 2¼ in. long, anchored by 5 in. square cast-iron washers and cotters. The base of the mortar is 9 in. thick and has a flange 3½ in. thick projecting 4⅝ in. From the bottom of the mortar (inside) to the lower projecting lip of the feed inlet is 28 in., this latter being 24 in. by 4½ in. wide, and from the mortar bottom to the lower and inner edge of the discharge aperture is 10 in., the breadth of the casting at this level being 13 in. There is a 2½ in. cast-iron false bottom and the four sides of the mortar have 1 in. steel liners, those on the ends being 16 in. high and the others 10 in. high. Above the back liner is a $\frac{3}{16}$ in. amalgamated copper plate 10 in. deep, the chuck blocks also carrying amalgamated plates. The tops of the dies lie ¾ in. below the level of the inner lip of the mortar discharge aperture, and the screen, of about 600 mesh, is carried by a frame presenting two apertures each 2 ft. 1 in. long by 12¼ in. deep.

The dies have their centres 10 in. apart and have a base 9⅝ in. square by 1 in. thick carrying a body 8¾ in. diameter by 6 in. high. The steel stamp stems are 14 ft. 6 in. long by 3¾ in. diameter with a 5½ in. taper at each end reducing it to 3$\frac{9}{16}$ in. There are three keys to each of the cast steel tappets which latter are 13 in. long by 9⅜ in. diameter, whilst the stamp heads, also of cast steel, are 18 in. high by 8¾ in. diameter. The shoes have a butt 9 in. high by 8¾ in. diameter and a shank 5¼ in. high and 4¾ in. to 3¾ in. diameter.

Each set of ten stamps is driven by a separate cam shaft 6 in. diameter having its centre 6 ft. 2 in. from the top of the battery posts and carrying a wooden pulley 6 ft. diameter with a 16½ in. face for a 14½ in. belt. The bearings, of the open type, are 14½ in. long on the pulley post and 12½ in. long on the others, and the centre of the shaft is 5½ in. from that of the stems. The cams are of cast steel and are 32 in. across the toes with a 2½ in. face, the boss being 11 in. diameter and 5 in. long and fitted with Blanton's keys. The guides are of the single cap type and the Jack shafts are 3½ in. diameter.

The stamps are driven from one lay shaft running down the centre of the mill having its bearings on the bottom transverse tie beams of the ore bin. The shaft is from 6 in. to 9 in. diameter and carries driving pulleys 52½ in. diameter by 16½ in., each provided with a Seymour friction clutch. This lay shaft runs at 65 revolutions per minute, the cam shafts thus making 47·4 revolutions and the number of drops per minute being barely 95. The cam floor is 17 in. below the cam shaft centre, the back floor being 5 ft. 6 in. wide and the front one 6 ft. 10 in., whilst the feeder floor stands 2 ft. above the level of the top of the mortar blocks. The ore is fed by suspended challenge machines and water is supplied by two 12 in. mains.

The amalgamated tables are 12 ft. long by 4 ft. 9⅜ in. broad and are set with a fall of 1¼ in. to the foot. They rest at the head on deals which bear on the upper supporting joist of the mortar blocks, and the foot is carried by an 8 by 4 in. deal supported by 6 by 6 in. posts. The heads of the table are 2 ft. 11 in. from the floor but running along and level with their feet, between them and the mill wall is a raised floor 5 ft. 6 in. wide.

The amalgamated surface consists of one plate $\frac{3}{16}$ in. thick bedded direct on the frame. Crossing the bottom of each two tables is a sloping launder which leads the pulp to a mercury trap from which it passes to a main launder running to the tailings wheel.

The ore bin, situated between the two rows of stamps, is 18 ft. wide with an A bottom, the depth on the centre line being 8 ft. 9 in. and at the sides 17 ft. 6 in. Its foundation posts, placed at the sides of the bin and behind each of the battery posts, are of 18 in. by 10 in. section and rest on mud sills 18 in. by 14 in. these being carried by the masonry steps already described. Each two foundation posts have a 12 by 12 in. tie beam in alignment with the battery line sills, these tie beams serving to carry the lay shaft. On these foundation posts rest the bin transverse sills of 12 by 10 in. section and in the frame formed by each of these sills the tie beam and the foundation posts are two 12 by 10 in. braces. The ore bin posts, of 12 by 10 in. section, rest on the transverse bin sills over each foundation post and are tied together across the bin by two 10 by 10 in. tie beams one at the top of the bin and the other at the level of the central apex, whilst two longitudinal cap sills, 12 by 8 in., cross these and are secured by 1⅛ in. bolts to the mud sills.

On each transverse bin sill rest the two diagonal beams of 10 by 10 in. section which carry the bin bottom each of which is stiffened at the centre by a 10 by 8 in. strut. Between these 10 by 10 in. bottom timbers are intermediate 10 by 5 in. joists. The sides and bottom of the bin are of 2 by 12 in. deals with a lining of 1½ in. boards. Over the bin run two tram tracks carried on 9 by 6 in. runners resting on 8 by 9 in. posts the feet of these bearing on the upper 10 by 10 in. transverse tie beams of the bin.

The clean-up appliances of the mill include a batea and a revolving drum.

The mill engine is a Cooper, compound tandem, surface-condensing horizontal, with 24 in. and 44 in. cylinders and 54 in. stroke ; the auxiliary is a Fraser and Chalmers' horizontal high-pressure engine with a 28 in. cylinder and 54 in. stroke.

THE CYANIDE PLANT.

The mill, cyanide works, and slime plant form three sides of a rectangle, the cyanide plant adjoining the mill, the leaching vats being arranged in two parallel rows at right angles to the mill and having the intermediate vats directly over them. From the centre of the mill to the centre of the near vats is 150 ft. and from these to the centre of the end vats is 215 ft., the distance between the vats in this direction being 43 ft. whilst from centre to centre of the vats rows it is 46 ft. From the centre of the end vats to the centre of the nearest slime vats is 115 ft., the longer axis of the slime plant running parallel with that of the mill.

The fall from the centre of the mill shaft to the top of the near end of the masonry walls which carry the cyanide vats is 7 ft. 6 in., whilst to the top of the walls of the end vats is 10 ft., the vats being stepped down. The fall from the centre of the mill shaft to the top of the slimes vats is 12 ft. and to the top of the solution settlers 24 ft. 3 in.

The mill launder delivers the pulp to a tailings wheel placed on the centre line of the sand plant with the centre of its shaft 16 ft. 5 in. above that of the mill shaft and 45 ft. from the centre of the nearest vats. The wheel is a double one carrying the belt between its two sets of buckets. Its diameter is 46 ft. and its breadth over all is 5 ft. 9 in., the shaft being 9 in. diameter with its bearings, which are hardwood blocks 14 by 16 in. by 3 ft. 3 in. long, 9 ft. 9 in. between centres. Each of these bearing blocks is bolted to a 12 by 16 in. beam anchored to a masonry pillar by two bolts 12 ft. long. These pillars are 27 ft. 5 in. high with a top section of 3 ft. by 6 ft. 6 in. and a bottom one of 20 ft. by 8 ft., the two pillars resting on a common base 22 ft. 6 in. by 20 ft. by 18 in. deep, the whole being built of good square hammer-dressed stone set in cement.

Jumpers Deep Gold Mining Company, Limited.

46 ft. Tailings Wheel, with Vats of the Cyanide Plant on the right.

I

The wheel has sixteen built arms each consisting of two 3 by 6 in. radial timbers framed with three 3 by 6 in. timbers and having six bracing bolts ⅝ in. diameter. Each radial timber is secured by two ¾ in. bolts to a casting 5 ft. diameter keyed to the shaft, there being two of these castings, one near each bearing, the distance between the radial timbers at the castings being 8 ft. 6 in. and at the periphery of the wheel 20 in. Between the radial timbers of adjoining arms are 4 by 6 in. ties, four between each pair of arms, those ties nearest the centre of the wheel having a ¾ in. bolt to a casting on the shaft whilst the outer ones are set just below the belt run. Between the outer ends of the radial timbers is the belt run 20 in. wide this being constructed of cross boards 4½ by 2 in, spaced 2 in. apart, resting on similar boards, these latter being carried on the edge of four 1½ in. boards cut to a circle 38 ft. 4 in. diameter.

The buckets, 256 in number—128 on each side of the belt run—are 12 in. deep and 15 in. wide and are tangent to a circle 33 ft. 2 in diameter, the pitch of their delivery lips being a little over 12 in. They are secured to the radial arms by ⅝ in. and ¾ in. bolts and 3 by 3 by ½ in. angle irons. Each set of 128 buckets delivers into a launder projecting under them the bottom of which is 15 ft. 9 in. above the centre of the wheel at one side and 14 ft. 10 in. at the other, ceiling boards 6 by ½ in. protecting the belt run from splash. These two launders converge and unite into one 2 ft. wide with a 3¼ per cent. fall which fans out at the end for the introduction of an automatic sampler, and thence the pulp passes to the concentrator boxes—four spitzluten—placed above and between the four intermediate concentrate vats. These boxes are of different dimensions the first being 3 ft. 8 in. deep from the inflow lip, the second 3 ft. 7 in., the third 3 ft. 6 in., and the fourth 3 ft. 5 in., but the long sides of all the boxes are 5 ft. 7 in. deep. The box centres are 4 ft. 3 in. apart and the areas at the inflow lip level are, first box 3 ft. 3 in. by 20 in., second box 3 ft. 3 in. by 2 ft., and the third and fourth boxes each 3 ft. 3 in. by 3 ft. The long sides of the boxes are vertical except for a few inches at the bottom where they take a slope of 45° tangent to the 3 in. discharge hole, the other sides being also tangent to this, the four sides meeting the base in a 3 in. square. The baffles are of 1 in. board and have their front face 4½ in before the box centres. The water supply for these concentrators comes from a rectangular tank standing over them with its bottom 11 ft. 6 in. above their base, this tank being 7 ft. 6 in. by 2 ft. 6 in. by 2 ft. 6 in. deep and from it goes a separate 2½ in. pipe to the bottom of each concentrator. The concentrates pass from the bottom of the spitzluten by pipes and launders to the intermediate concentrate vats whilst the sand and slime flow to the spitzkasten placed in front and above the first pair of intermediate sand vats. The pulp, which in passing from the concentrator boxes encounters another automatic sampler, enters the spitzkasten by a large fan launder 4 ft. wide at the inlet and 25 ft. 6 in. broad at the delivery into the spitzkasten, this having its long axis across that of the plant and consisting of a box 25 ft. 6 in. long by 6 ft. 3 in. broad divided into four pyramidal compartments each 6 ft. 3 in. square at the top and with a base 9 in. by 9 in. the depth of the boxes being 5 ft. 3 in. To the bottom of each compartment is a 2 in. branch pipe from a 3 in. main connected with the water tank over the concentrator boxes, and the sands leave each division of the spitzkasten by a 3½ in. pipe which runs the length of the plant with a 3·4 per cent. grade, the pulp being passed from these to the vats by hose pipes. The slime proceeds from the lip of the spitzkasten to a second spitzkasten similar to it, but only half the size, which settles any sand in suspension and returns it to the wheel.

There are twenty-four vats—twelve intermediate vats and twelve leaching vats—the former being supported directly over the latter, the distance between the bottom of the upper vats and the rim of the lower ones being 5 ft. ⅜ in., the plan of the plant showing two rows each of six vats. Eight of these vats—four intermediates and four leaching vats—are reserved for concentrates.

The intermediate vats are of steel and are 40 ft. diameter by 7 ft. deep internally, the sides being ¼ in. thick and one plate deep, the vertical joints being single rivetted lap joints with ½ in. rivets having a pitch of 1¾ in. and an overlap of 1⅛ in. The rim of each vat is stiffened by an external angle iron 3 by 3 by ⅜ in., a similar external angle iron securing the sides to the bottom. The latter is of ₅⁄₁₆ in. plates with parallel single rivetted lap joints. Each bottom has seven 20 in. discharge doors of plug pattern and each vat has three side discharge gates 3 ft. wide, slat type with roller curtain, for the water and slime which a short pipe conveys to a slime launder running to the second spitzkasten. These intermediate vats have the ordinary filter bottom.

The twelve leaching vats immediately beneath are 37 ft. in diameter and 10 ft. deep, the shells being similar in construction to those of the intermediates except that the overlap of the joints is 1½ in. instead of 1⅛ in. and they each have eight bottom discharge doors 23½ in. diameter.

The bottoms of the leaching vats rest on 9 by 3 in. deals lying flat with 3 in. spaces and crossing 8 by 4 in. rolled steel joists (25 lbs. per foot) of which there are sixteen under each vat of lengths varying from 12 ft. 6 in. to 37 ft. 6 in., these bearing on the masonry piers which are at right angles to them, running the long way of the plant.

There are four piers under each of the two rows of leaching vats, forming three discharge ways. The two inner piers of each row are continuous, but between the vats they step down 6 in., the difference in level of the ends of the piers being 2 ft. 6 in. The side piers are not continuous but have a length of 31 ft. 2 in. under each vat. The masonry is 3 ft. 3 in. thick at the top and 4 ft. 3 in. at the bottom and is of large squared stone set in lime mortar except at four points—two on each of the inner piers—under each vat where they are expanded to form pillars 4 ft. 6 in. square at the top by 5 ft. 6 in. square at the bottom and are set in cement mortar, each of these pillars supporting one of the cast-iron columns which carry the upper vat. The piers are about 6 ft. high and from the centre of the vat to the centre of the inner ones is 6 ft. 2 in. and to the outer ones 18 ft. 6 in.

The intermediate vats are each carried by twelve cast-iron columns—four inside the lower vat and eight outside—the first bearing on the vat bottom over the masonry pillars mentioned above and the latter resting on the masonry piers just outside the lower vats, two on each of the four piers. The inside columns are 11 in. diameter and of 1 in. metal and have a length of 12 ft. ⅜ in. with a top flange 16 in. square and 2 in. thick and a bottom one 2 ft. 9 in. square of similar thickness. The distance from centre to centre of these along the piers is 15 ft. 2 in. Between their bottom flange and the vat bottom is 1¼ in. of cement grouting and between the vat bottom and the masonry pillar below there is a concrete block. The two columns on each of the outside piers are 13 ft. 3 11/16 in. long and 7 in. diameter with 1⅛ in. metal and have a top flange 12 by 12 by 1¼ in. and a bottom one 21 by 21 by 1¼ in.; they also each have two side brackets to carry flooring joists. The two outside columns on each of the inner piers are 13 ft. 1 7/16 in. long and 7 in. diameter with ⅞ in. metal and have a 12 by 12 by 1¼ in. top flange and a bottom one 21 by 21 by 1¼ in. Four 1⅛ in. bolts 18 in. long anchor each of the outside pillars to the masonry piers.

The four columns on each inner pier carry a main steel girder 39 ft. long of 19¼ in. by 12 in. section (130 lbs. to the foot) and the two columns on each of the outer piers carry main girders 17 ft. long of 17 by 10 in. section (98 lbs. per foot). Between these run diagonal corner girders 17 ft. 3 in. long of 14 by 6 in. section (57 lbs. per foot). On this girder frame rest the cross joists which carry the 9 by 3 in. deals supporting the intermediate vat bottoms, these deals being arranged like those of the lower vats. These steel cross joists, of which there are fourteen under each vat, are of 12 by 5 in. section (32 lbs. to the foot) and are of lengths varying from 19 ft. to 40 ft. 6 in. The columns of adjoining vats are tied together by 12 by 5 in. and 8 by 5 in. steel joists resting on their brackets, these joists carrying the lower platform.

Two 5 in. pipes running the length of the plant, one between the intermediate vats and the other between the leaching vats, supply the solutions to each vat through a 5 in. branch. For draining water from the intermediate vats each of the latter has a 2½ in. branch running to a 3 in. main whilst for solution leaching these vats have each a 2 in. pipe to the extractor house. Each of the lower vats has an independent 2½ in. leaching pipe.

The lower vats are discharged into trucks running in the passages below them and the sand is taken by endless rope haulage to the dump.

THE SLIME PLANT.

This consists of fifteen agitator vats, arranged in three parallel rows of five vats each, and two solution settlers, the rims of the former being 2 ft. below the top of the near end of the masonry piers of the sands plant and the top of the solution settlers 14 ft. 3 in. below the same point. The agitator vats, to which the slime comes direct from the second spitzkasten of the sand plant, have their centres 43 ft. apart the long way of the plant and 46 ft. the short way and the vats, which are of steel, are 40 ft. diameter by 7 ft. 6 in. deep on the one side and 8 ft. 6 in. deep on the other, their tops being horizontal but the bottoms having a fall of 12 in., ten of the vats—two rows—sloping to a cemented trench running between them which carries the piping, the other row being inclined to a separate but similar trench passing in front of them. The bottoms of the vats rest on well-tarred 9 by 3 in. deals, spaced 2 in. apart, resting on a foundation of broken stones rolled hard and flat.

The pipe-carrying trenches already referred to, the sides of which are of rubble masonry set in lime, are 3 ft. wide at the bottom which is covered with 3 in. of concrete slightly hollowed to the centre and having a fall of 2 per cent., the least depth being 18 in.

The two trenches have a small recess 3 ft. 3 in. by 3 ft. under the front of each vat for the discharge pipe elbow and they are connected by a cross trench with the extractor house. Down and over the centre line of each row of five vats runs the supporting gear for the agitators this consisting of two parallel channel irons 12 by 3½ in., placed 10¼ in. apart, each trussed with a 4 by 4 by ½ in. angle iron and having a 4 by 4 by ¼ in. T iron to the rim of the vat to carry a platform near the centre of the latter. The channel irons bear on the bottom flange of distance castings which rest on the rims of the vats, and between each pair of vats they are also supported by an iron standard 8 ft. 1½ in. high constructed of 4 by 4 by ½ in. angle iron its base being anchored to a masonry foundation by three 1 in. bolts.

The channel irons have brackets for carrying the 4 in. line shafts which drive the agitators, there being three of these shafts, one to each row of five vats. The agitator of each vat consists of a central hanging vertical shaft 6 in. diameter running on a ball bearing (29 1½ in.

balls) carried by the channel irons. Its lower end enters a footstep fixed on the vat bottom and lined with lignum vitæ. Twelve inches from the vat bottom the shaft has secured to it a cast-iron boss to which are attached the six radial T iron arms of 4 by 5 in. section, these arms being equally spaced and approaching to within 12 in. of the vat side. They have a slight fall from the boss and are each supported by a ¾ in. bolt stay running to a casting fixed on the shaft about level with the rim of the vat. To the top of this hanging shaft is keyed a worm wheel having thirty teeth of 3 in. pitch which gears with a worm on the line shaft, a claw clutch throwing each worm in and out of gear. The line shafts make 120 revolutions per minute and the agitators four per minute the former being driven by a high-speed rope from the mill, an electric motor in the extractor house being pressed into service in the event of accident to or stoppage of the rope drive.

Each vat has three exit pipes and two supply pipes. Of the former, one is the 6 in. sludge discharge pipe attached to the bottom at the lowest point and connecting with the 9 in. sludge main running to the discharge pump in the extractor house which transfers the treated slime from the vats to the dam. The other two exit pipes are the 5 in. water decanting pipe and the 3 in. solution decanter which are fixed on the side of the vat near the bottom ; each of these project into the vat in the form of a swivel pipe the end of which can be raised and lowered in a vertical plane. Of the supply pipes one is a 5 in. solution pipe and the other a 5 in. pipe for supplying water or transferring sludge from vat to vat and, in case of accident to the agitators, when the pulp is kept in motion by pumping, it serves as the return slime pipe.

The two solution settlers of this slimes plant through which the solutions pass before going to the extractor house are each 31 ft. 6 in. diameter by 9 ft. deep and are of steel. Their centres are 38 ft. apart and 58 ft. from the centre of the near row of agitator vats whilst their level is such that the solution from the latter gravitates to them.

The extractor house is 160 ft. long by 60 ft. wide and has its sloping floor 11 ft. 4 in. to 12 ft. 4 in. below the top of the near end of the masonry piers of the sand vats. Here the solutions of both the sand plant and slime plant are precipitated, the former by the zinc process and the latter by the electrical method. The solutions from the sand plant gravitate to a series of six launders in the house four of which deliver to steady head vats, one to each, the other two going to the concentrate precipitation boxes. There are seven zinc boxes in all, two for concentrates, 19 ft. by 3 ft. by 2 ft. 6 in. deep, and five for sands, 19 ft. by 5 ft. by 2 ft. 6 in. deep, each box having eight zinc compartments and a 3 in. inlet and outlet pipe.

The electric precipitation boxes for the solutions from the slime plant are six in number and are 40 ft. by 6 ft. by 5 ft. deep each with nine compartments. Strip lead cathodes are used and the inlet pipes are 3 in. branches from 4 in. mains which run from steady head boxes into which the solutions are pumped from the two solution settlers. Wood launders convey the solutions from these precipitation boxes to the sumps.

The machinery in the extractor house comprises :—

One Riedler pump for returning water from the sludge dam.

One 5 in. centrifugal sludge pump for transferring the sludge from the vats to the dam.

One 5 in. centrifugal pump for transferring the slime from tank to tank and for agitating the charge in case of breakdown of the ordinary gear.

One 5 in. centrifugal pump for supplying water to the slimes vats.

One 5 in. centrifugal pump for supplying solution to the slimes vats.

One 5 in. centrifugal pump for supplying solution to the sand plant.

One 3 in. centrifugal for transferring solution from the two solution settlers to the steady head boxes feeding the electrical precipitation boxes.

One electric motor for the pumps and also for driving the slime agitators when the rope drive breaks down or stops.

One plating dynamo for the Siemens-Halske precipitating boxes.

One Tangye 12 by 6 in. single-acting vertical vacuum pump with intermediate chamber.

One filter press for gold slimes.

One Ball mill for slags.

Two turning lathes for zinc.

The clean-up apparatus includes :—

Two zinc calciners, two amalgam retorts, one lead melting furnace and four fusion furnaces.

Outside the extractor house are the eight sump vats of which two are for the slime plant and two each for the strong, medium, and week solution respectively of the sand plant. They are 31 ft. 6 in. diameter and 9 ft. 6 in. deep and are arranged in two parallel rows resting on the bottom of an excavation.

Of the three plants described that of the Jumpers Deep is the latest and represents the most advanced ideas. Under the management of Mr. G. E. Webber and Mr. S. Hancock and with its engineering in the hands of Mr. L. J. Seymour much was expected, and, judging from the plans and the plant already erected, the installation will be one that will command admiration for its arrangement and engineering.

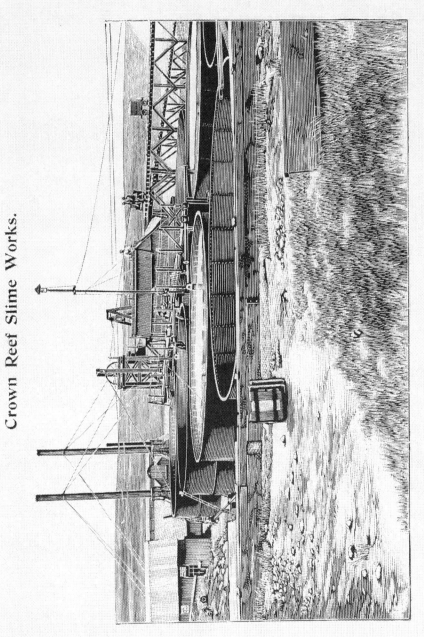

Crown Reef Slime Works.

Showing the two Intermediate Vats, the six Agitator Vats. and the Extractor House.

CROWN REEF GOLD MINING COMPANY, LIMITED.

SLIME PLANT.

THIS slime works, the second to commence operations on these fields, is attracting much attention by its satisfactory commercial results and its somewhat novel design, a design which differs materially from that of the Jumpers Deep plant although the process is the same, the current slime being treated by agitation in weak cyanide solutions with subsequent decantation of the clear auriferous liquor. The following is an outline of the plant and the method of working :—

The slime from the spitzkasten feeding the sand vats and that overflowing from these vats is carried by a launder to three excavated settling pits lined with stone, lime water being added to the pulp in the launder to expedite the settling. Of these settling pits, which are of pyramidal shape each of them tapering to a point at the bottom, the first two have each a top section of 20 ft. by 20 ft., the third one being 40 ft. by 40 ft., the depth in each case being about 10 ft. They are arranged side by side and the slime is divided and fed into the first two pits these being each provided with two parallel surface launders which take the overflow and convey it to a cross launder at the foot of the pits, the latter launder carrying the overflow to the third pit from which the now clear water escapes both at the side and into a launder running diagonally across the surface of the pit.

The slime settles to the apex of each pit from where it is drawn by a centrifugal pump and delivered into one of two intermediate vats, 32 ft. diameter by 10 ft. deep, having iron sides and a concrete bottom which slopes to a central discharge aperture. Here the slime settles and the water is decanted off at the side and returned to the settling pits, after which cyanide solution

K 2

of about ·008 per cent. strength is pumped in and this and the slime are simultaneously transferred by pump to one of six agitation vats, 32 ft. diameter by 10 ft. deep, into which 80 tons of similar solution have already been run. Each of these vats takes a charge of 60 tons of slime, 120 tons of cyanide solution coming with this from the intermediate vats, the total charge being thus 60 tons of slime and 200 tons of solution. These agitation vats have concrete bottoms sloping to a central discharge aperture like the intermediate vats, and the agitation of the charge is effected by a centrifugal pump which sucks the pulp through a pipe connected with each of these apertures and returns it by other pipes running near the bottom of the vats into which they have movable branches projecting.

After about 1½ hours agitation the charge is allowed to settle and the clear solution is decanted off and run to two solution settling vats 20 ft. diameter, whilst the sludge is discharged to the dam. Each charge is in the vats about thirteen hours. From the solution settlers the liquor gravitates to four electrical precipitation boxes in the extractor house, each of these boxes being 30 ft. by 4 ft. by 6 ft., and passes thence to a large sump vat, 40 ft. diameter, having a capacity of about 70,000 gallons.

For the necessary pumping there are four 4 in. centrifugals placed in a pit at the head of the extractor boxes and driven from a shaft run by an electric motor giving 500 volts 40 amperes at 800 revolutions, this also driving the dynamo which furnishes the current for the precipitation boxes. Of these pumps one is for the solutions, one for transferring the slime from the settling pits to the intermediate vats, another for agitating the charges, and the fourth for discharging the sludge, the pipes being so arranged that the duties of the pumps are interchangeable. A 6 in. centrifugal pump has just been added for agitating purposes.

The solutions which enter the extractor boxes carrying about 30 grains of gold per ton leave them with about 6 grains per ton and the theoretical extraction on the slime is claimed to be about 80 per cent. and the value of the residue below a pennyweight per ton. The cost of treatment is now averaging about 4s. 8d. per ton exclusive of depreciation of plant.

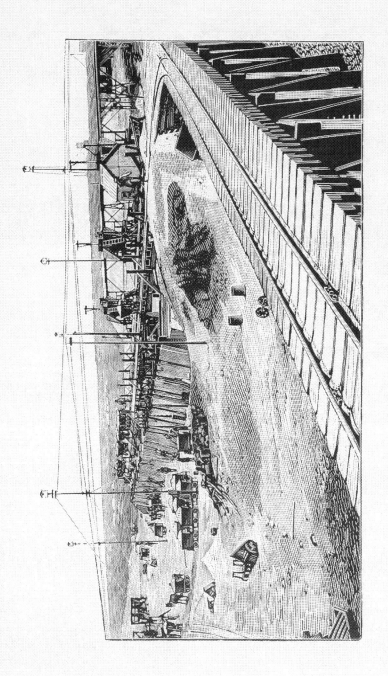

The Crown Reef Cyanide Works.

GENERAL REVIEW OF RAND PRACTICE.

HOISTING.

THE number of main shafts on the various properties ranges from one to eight, the Simmer and Jack Proprietary Mines having the latter number. Most of the deep level mines have two, each about 22 ft. by 6 ft. in the clear. With the exception of the circular shafts at the New Primrose, United, and Langlaagte Royal Mines, they are of rectangular section and of sizes varying from about 11 ft. by 5 ft. to 26 ft. by 6 ft. with from one to four hoisting compartments; the long axis of the shaft generally lying parallel with the strike of the reef, the shafts at the New Primrose being among the few exceptions to this rule. On the deep level mines the shafts are vertical down to their intersection of the reefs and from there proceed on the underlay, the depth of the vertical section being within the wide limits of 350 ft. and 3,300 ft. Some have independent winding arrangements for the two sections, but others, like the Jumpers Deep, have a single system. The majority of the shafts of the outcrop companies are inclined throughout but some few intersect the reefs vertically and are then continued as inclines like the deep level shafts.

With shafts having two or more winding compartments it is customary to reserve one or more for development and general work, these, as a rule, being provided with smaller skips and an independent engine. In the inclined shafts automatic tipping skips of capacities ranging up to 3 tons are in general use and their satisfactory and expeditious working is leading to their adoption in the vertical shafts.

Rand winding engines include both the geared and direct-acting pattern, but the increasing depth of the workings and the higher speeds now required are creating a growing demand for the latter type.

One of the striking features of this gold field is its great headgears which are now invariably constructed of timber and are of heights rising from 40 ft. to 88 ft., the Knight's Central Mine possessing a gear of the latter size. In the early days a few steel headgears were erected and some are still in use, the George Goch having one of 50 ft., the Jumpers one of 34 ft., and the Langlaagte Estate one of 35 ft. As a constructive material for the gears steel would be more used if it were not for its greater cost, for with wood there is the objection that the great length of leg called for necessitates each being built of several timbers and trouble occasionally arises from chafing at the joints. Many of the headgears are provided with grizzlies and several ore bins and it is common practice for them to carry the sorting and breaking plant which, when not so installed, is usually to be found in a separate sorting and crushing station, there being very few instances of breakers being located in the mill itself.

SORTING.

Of late much attention has been devoted to the question of close sorting above ground and its value is becoming recognised and the practice widely adopted, but there are still mines, like the Lancaster, where no provision is made for it. The percentage sorted naturally varies inversely with the size of the reefs, the maximum amount sorted out at any of the mines being about 50 per cent. of the rock raised.

Whilst it is generally agreed that where there is only one main shaft the crushers are best placed in its headgear, there are various opinions held regarding the best position for the sorting and crushing plant on properties having more than one main shaft, but recent installations have discovered a leaning in favour of an independent crushing station between the shafts and the mill. This arrangement, compared with that of several independent installations in headgears, has the advantage of requiring less expenditure for headgear and plant and is accompanied by smaller working costs and a more efficient sorting, but unless such a central station has ample auxiliary plant there is more danger of stoppages and delays attending it.

For sorting we have stationary floors as at the Ferreira, bumping tables as at the Roodepoort United Main Reef, travelling belts as at the Meyer and Charlton, and revolving circular tables as used at the Robinson. Of these the travelling belts and circular tables are preferred and the latter give most satisfaction because of the less wear and tear attending their running, but the wide floor they require is somewhat of an objection to them. In the Langlaagte Royal main headgear there are two such tables, each 24 ft. diameter. An innovation in connection with sorting at the Wolhuter Mine deserves notice. Here Mr. Britten, the general manager, has two grizzlies—one coarse and one fine—arranged one above the other in order to obtain two products for sorting, the idea underlying the practice being that if the coarse and fine ore are sorted together the lumps cover and hide the latter to some extent and so interfere with satisfactory sorting.

Regarding the construction of circular sorting tables it is well to point out that central vertical shafts are being abandoned in favour of supporting the tables on a rail and flanged wheels.

BREAKING.

Turning to crushers we have Blake machines as at the Treasury and Lancaster, Gates as at the Robinson, Comets at the Jumpers and Crown Deep, and Wells' breakers at the New Primrose and Glencairn Mines. To their comparative freedom from vibration and their great capacity the gyratory crushers owe their extensive adoption here, for these qualities were such as were desired in machines for the high headgears and central stations. Blake machines are, nevertheless, doing excellent work in places and their wear and tear is much less than that of the Gates type of which the repairs and renewals are also of a far more troublesome nature. Fine crushing in the breaker station is growing in favour and it is not unlikely that the practice of two-stage breaking, with a view to obtain a small-sized product, will gain ground. The Langlaagte Royal has two-stage breaking, the main shaft headgear having two No. 5 Gates feeding two No. 3's, a shaking screen intervening between the two sets of crushers. The Wells' machine mentioned is a double-ended reciprocating crusher of the Blake type, the two moving jaws being carried on one solid casting which is recessed for the sliding block of the eccentric which actuates it. Here it may conveniently be mentioned that there are many engineers on these fields who advocate, on economic grounds, the amalgamation of the first row of deep level properties with the outcrop companies, and the great saving in shafts and in surface equipment, together with the lower working costs which might reasonably be expected to result from such amalgamation, are weighty arguments in its favour.

The methods of transporting the ore from the shafts to the mill are many, the contour of the property and the number of the shafts and their position relatively to the mill determining the practice adopted. At the Witwatersrand Mine native labour is employed; at the

L

Geldenhuis Estate and many other mines endless rope haulage is in use; the Jumpers Deep will have a single-rope system from the crushing station to the mill and endless rope from shafts to crusher station; whilst at the Salisbury and Jubilee Mines mules do the work. Besides these agents we have electric locomotives at the Crown Reef and Langlaagte Royal, and steam locomotives at the Lancaster and the Simmer and Jack Proprietary Mines.

The trucks usually reach the top of the mill ore bin by an inclined plane, but at some mines there are vertical hoists for raising the trucks to the bin floor, whilst at the Geldenhuis Estate neither hoist nor inclined plane is needed the trucks running in on the level. Inclined planes to the bin floor are little trouble but require much space ; vertical hoists take up less room but need more labour.

BATTERIES.

The position of the mill depends on the number and position of the shafts and other circumstances but with one main shaft the mill is usually placed as near to it as conditions permit, whilst with more than one the selection of site is influenced by the facilities afforded communication with the shafts and the advantages offered for the cyanide treatment, sand dump and the conservation of slime and water ; the position of the water supply is but a secondary consideration. The Geldenhuis Estate is one of the very few mines on the Rand with a good natural mill site and its ore is trammed 3,600 ft. to take advantage of it.

Coming to mill practice, the large number of stamps on many of the properties naturally led to arranging them in two rows in order to keep the mills compact and the cost of erection down. The City and Suburban and the Geldenhuis Deep are examples of back to back mills, whilst the Wolhuter and the Witwatersrand are types of single-line mills, the 100 stamps of the Wolhuter being run from a shaft with end drive, whilst the lay shaft driving the 120 stamps of the Witwatersrand is centre driven.

For some time past the efforts of Rand engineers have been concentrated on increasing the stamp duty, milling results indicating that very fine crushing was uncalled for and that a good amalgamation inside the mortar was not generally essential to satisfactory extractions, and, therefore, that the stamps might with advantage be treated and perfected as crushing agents pure and simple. Their development has taken the direction of having a low discharge of about 4 in. with screens of 600 to 900 mesh and a width of mortar at the issue level of about 12 in., whilst the weight of the stamps has been gradually increased until it has reached 1,250 lbs., the drops being 90 to 100 per minute with 7 to 8 in. fall. Mr. Blane, of the Glencairn, who has paid much attention to this matter, prefers a height of discharge of 3 in.,

The Simmer and Jack Proprietary Mines, Limited.

280 Stamp Mill.—Showing Amalgamated Plates in Position.

having proved by experiment that this gives a high duty without undue wear of screen. The 1,250 lbs. stamps of the New Modderfontein are giving a good account of themselves, crushing over 6 tons each per 24 hours, but, notwithstanding this, the opinion that the limit of weight might advantageously be placed at 1,050 lbs. frequently finds expression.

The use of liners and false bottoms in mortars is becoming more general, whilst the poor results given by internal amalgamated plates is creating a disfavour for them. As a substitute for these plates the New Crœsus and Simmer and Jack mortars have cast steel recessed liners which have not been fully tested here as yet but are in use at the Alaska Treadwell Mine. At the New Primrose Mine internal amalgamated plates have been discarded and it is not unlikely that the practice of adding mercury in the boxes will also be discontinued at this mine, it being asserted that although with oxidised ore the addition of mercury in the box may be advantageous, with " blue " or unoxidised banket the mercury " flours " too much, causing a loss of both mercury and gold. There are few lip and splash plates in use and no mercury wells, whilst stepped tables are uncommon it being customary to have the amalgamated surface in one plate.

The stamps, which are invariably in batteries of five, are usually arranged in sets of ten with a space between the sets and a cam shaft to each five stamps, among the exceptions to this rule being the Simmer and Jack Proprietary mill in which the stamps are erected in sets of twenty, and the Ferreira mill where the stamps have a space between each battery of five. There are advocates for both five and ten cams on a shaft, and among the various opinions on this point is that of Mr. J. B. Roberts, consulting mechanical engineer, who has pointed out that with ten cams on a shaft there must be greater danger of it fracturing than when carrying five because of the increased molecular change in the metal of the shaft induced by the greater number of shocks it is subjected to. Mr. Roberts also suggests that to diminish the torsion on a ten cam shaft and the twist on its battery posts the pulley might be placed on it at mid length instead of at the end as is usual.

Rand battery frames are of the reversed knee pattern, and the mortar blocks, which frequently reach a length of 15 ft., are generally bedded on concrete. The mortar blocks of the Geldenhuis Deep are of concrete, the ground having been judged too yielding to offer the requisite resistance to the small area presented by ordinary mortar blocks. Large concrete blocks, 6 ft by 5 ft. 6 in. by 11 ft. 6 in., were put down, one for each battery of ten stamps, and the mortars are bedded on frames of hardwood resting on these. It is questionable whether ordinary mortar blocks on massive concrete foundations would not have been a better arrangement, for the tendency of concrete to disintegrate when subjected to constant jar and vibration points to the advisability of interposing some elastic buffer between it and the mortar. Mr. Price, the present general manager, has recognised this call for elasticity, and in the case of a few of the stamps has improved on the original arrangement by placing a pitch pine frame between the 6 in. cast-iron sole plates upon which the mortars bed and the concrete blocks. He has also effected an improvement in connection with the holding-down bolts which he has made accessible to permit of renewal in case of fracture.

The motive power of Rand mills is steam and there are many fine engines, the horizontal type being preferred to the vertical class. Of vertical engines there are good examples at the Geldenhuis Estate and City and Suburban, that of the latter being tandem, cross-coupled, triple-expansion and condensing, with Corliss valve gear, the cylinders being 20 in., 31 in., and 48 in. diameter, with a 4 ft. stroke, whilst that of the Geldenhuis Estate is a 750 h.p. compound vertical by Yates and Thom. The Robinson Mill has a vertical engine of the King type with a high-pressure cylinder 19 in. diameter, an intermediate one of 30 in. diameter and two low-pressure cylinders of 30 in. diameter, the stroke being 42 in. Of horizontal mill motors, the engines of the Simmer and Jack, Glencairn, and Jumpers Deep are good examples. With two lay shafts rope drives are superseding belts, but where there is only one lay shaft the engine frequently drives direct, the shaft being a continuation of the engine crank-shaft as at the Robinson and Ferreira. It is now found expedient for the mill engine to drive, besides the battery, the dynamos for providing power for most of the surface machinery, and also to run the electric lighting plant. Opinions regarding the wisdom of having auxiliary mill engines are very conflicting. Thus Mr. Way, general manager of the George Goch Amalgamated Company, one of the Albu group of mines, believes in having the mill engine in duplicate, whilst others consider that an auxiliary engine capable of driving all plant but the mill satisfies requirements, contending that to duplicate the mill motor is very expensive and that a well-designed engine is little likely to go wrong or cause delay if thoroughly overhauled each month. In concluding this review of mill practice it may be mentioned that an innovation has just been effected at the May Consolidated Mine where the mill is being driven by electricity supplied by an Electric Supply Company which has its central station near the Brakpan Collieries. The experiment will be watched with interest in view of the high freight charges on coal in the Transvaal.

CYANIDATION.

There are few mines which do not separate a concentrate from their mill pulp, but the advance of the cyanide process has to a very great extent done away with the necessity for close concentration and the subsequent chlorination generally accompanying it. The

Intermediate Vats of the New Primrose Cyanide Works.

Robinson and Ferreira are among the few mines which still find it advisable to run vanners and chlorinate, and it is worthy of note that it has been decided to include vanners and a chlorination plant in the new installation of the Violet Consolidated Mines. The heavy cost of obtaining a close concentrate and the high charges attending chlorination have resulted in the present general practice of effecting a rough concentration in spitzluten and treating the concentrates by cyanide, to which solvent they frequently yield a high extraction; this method gives good general results on the majority of the ores, but some few appear to give better returns with chlorination of vanner concentrates and cyanidation of the impoverished sands. At the George Goch Mines vanners are to be used and the coarse concentrates will be cyanided whilst those from the slime are to be treated by chlorination. The pulp of each amalgamated table will pass through a spitzluten where lime is added, and from the bottom of this the coarse sands are to be led to two Frue vanners whilst the overflow—the slime—after going through a second spitzluten is passed to another vanner where it is expected that the lime added will result in such a concentration and impoverishment as to render a slime plant unnecessary.

The configuration of the Rand is such that the mines, with very few exceptions, have to elevate the mill pulp to obtain the fall necessary for the subsequent treatment. For this purpose bucket-wheels are used wherever possible, but when the lift calls for a wheel over 50 ft. diameter their cost becomes prohibitive and the coarse sand has a tendency to settle in the buckets; therefore, for such high lifts as this and higher ones, plunger pumps are used which are much more costly to keep in repair than the wheels which give little trouble. The Simmer and Jack tailings wheels, already described, are good examples of the single type, and that of the Jumpers Deep of the double class. Iron wheels and iron buckets in timber wheels have been tried, but, unless kept well tarred, metal buckets soon corrode and wear through. With ore carrying a high percentage of pyrites it has been found necessary to put fillets in the buckets to render their angle less acute and so prevent the settling of the pyrites. Tailings pumps are usually in pairs, like those of the Glencairn, and they are constructed with a separate delivery main for each plunger, experience having shown that with two plungers delivering into one rising main they do not work satisfactorily.

Turning to the question of concentration for the cyanide process, the aim is not to obtain a number of concentrated products of different grades but a somewhat bulky concentrate of one grade, so, although it is customary to use three or four spitzluten, the boxes of a series vary little in size and their products are combined.

M

Such a new and progressive process as cyanidation has naturally offered great scope in the construction and arrangement of its plants, and the works which have been erected are of diverse designs. Among the various types we have, first, the plants without intermediate vats, comprising (a) those with a single group of excavated and lined-treating vats, the sands being settled in pits or dams, as at Langlaagte Estate; (b) plants with a single group of treating vats which are arranged above ground level, the sand being settled in these vats, as at the Jumpers, or in pits. Of works with intermediate vats we have (c) those arranged with the vats in two groups at or about the same level, one lot being intermediate and others the leaching vats, an inclined plane carrying the sand from one set to the other, as at the Simmer and Jack; (d) plants with the intermediate and leaching vats arranged in two groups such that the level of the intermediates allows of the trucks running from beneath them to the leaching vats on a level plane, as at the New Comet and New Primrose; (e) those in which the intermediate group of vats is at a lower level than the leaching vats, with an inclined plane leading to the top of the latter, as at the Ferreira; (f) plants in which there is an intermediate vat supported directly over each leaching vat, as at the New Crœsus.

With the initiation of direct filling the cyanide process made a notable step forward, the utility and advantage of the idea being testified to by its general adoption. The practice is to run the pulp into so-called intermediate vats, from which, after settling and draining, the sand is taken to the leaching vats. Attempts to complete the treatment in the direct-filled vats have not proved altogether successful, but at the Jumpers Mine, where this is done, satisfactory results are being obtained; similarly the so-called double treatment, which consists of giving one or more weak cyanide washes· in the intermediate vats previous to transferring to the leaching vats, has not been in all cases an unqualified success.

The general approval of intermediate filling has related plants of (a) and (b) design to the obsolete types and called into existence others which afford special facilities for conveying the sand from vat to vat. Of these the superimposed arrangement (f) is deservedly being most adopted, for it requires the minimum labour in transferring and gives a very compact plant. It has been urged as an objection to plants built on these lines that the handling during transference is insufficient to prevent "lumps" reaching the leaching vats and that the (d) type is preferable on this account because the tramming between the vats breaks the lumps, but against this belief it is contended that a grating under the intermediate vats of arrangement (f) would suffice to destroy such "lumps" and that the vibration and jar experienced by the sand in the tramming associated with type (d) are such as to induce agglomeration rather than disintegration, especially if much moisture is present and the trucks are V-shaped. Plants of type (c) require less elevating of the mill pulp than the super-

New Crœsus Cyanide Works, showing Conglomerate Piers.

Witwatersrand Gold Mining Company, Limited.

View of Cyanide Works, showing Intermediate Vats supported on cast iron pillars and steel girders. Plant being extended.

imposed system and with reference to the (*d*) arrangement it would be more in evidence had ground favourable to it been more generally available. The May Consolidated and the New Comet plants are of this latter description, but their sites being flat the intermediate vats are carried on high timber structures and this adaptation of the system has not generally commended itself.

The construction of the vats is almost as varied as the types of plants. We have them of concrete, brick and cement, wood, iron and steel, and of iron with wood and concrete bottoms. Concrete and brick vats have become things of the past owing to their great cost and less convenient working, and wood as a constructive material is rapidly being superseded by iron and steel which have the great advantage of not suffering from exposure or from being alternately dry and wet; timber vats, especially if of Oregon pine, are much distressed by this. With iron vats, where the water is acid, it is advisable to tar the inside of the intermediates, but the leaching vats may be left bare as the cyanide has no appreciable destructive action on the metal or the metal on the cyanide solution. Vats with iron sides and timber bottoms, of which a few were erected, have disclosed no special merit, whilst the expansion and contraction of the iron when the vat is exposed has racked the bottom joint and induced leakage. Treating vats with cement bottoms and iron sides are used at New Modderfontein, the bottom being flush with the ground. When first introduced iron vats had bolted joints, but putting them together, although it does not require skilled labour, proved so tedious and troublesome that they are now rivetted, this practice also giving tighter joints. The vats of the Jumpers Deep are good examples of their kind, and the difference in the dimensions of the intermediate and leaching vats of this mine indicates the extent of the allowance generally made for the increased bulk of the sand when it leaves the intermediate vats.

The great capacity of present day leaching vats and the introduction of bottom discharge has effected a change in the method of supporting them, this now almost invariably taking the form of massive masonry or conglomerate piers of sufficient height to permit trucks to pass beneath the vats. The manner of supporting the intermediate vats varies, of course, with their arrangement, but when directly over the leaching vats they are carried either on timber framing, as at the Glencairn and the New Crœsus, or on steel girders and cast-iron columns, like the Witwatersrand plant and the Jumpers Deep; the latter method makes a first-class job but is more expensive. At the City and Suburban Mine, where formerly the treatment of the sand was completed in the direct-filled vats, iron intermediate vats have been placed over the old cement ones.

The distribution of the sand in the intermediate vats is generally effected through the medium of a hose pipe or automatic distributor, but the practice of running the pulp in at one side of the vat and having the overflow at the other side is not unknown, and there are cases where the pulp is led by a fixed launder to the centre of the vat and there discharged, this being the present manner of filling intermediate concentrate vats. Of late hose filling has gained ground and at the Robinson Mine automatic distributors have been discarded and hose resorted to, the reason being that here, as at other mines, the percentage of sand caught with automatic filling was too low, the hose giving a greater tonnage. That hose filling can be manipulated to give a high retention there can be no doubting, but the method is a very crude one and the results are dependent on the man at the hose—a Kaffir—whilst the sand cannot but be deposited in an irregular fashion. The practice, however, of having separate vats for the operations of direct filling and leaching has made an even distribution in the direct-filled vat less vital, for the sands are well mixed in discharging them from the one vat to the other, and as there is a growing desire to catch a high percentage of the pulp in the intermediate vats hose filling is coming more to the front. At the Crown Reef and City and Suburban Mines, where this method of filling is in use, it gives satisfaction. But although automatic distributors have fallen behind of late their principle is excellent and their regular working and independence of supervision is much appreciated. Their alleged failing, the loss of sand, is due not so much to the distributor as to the position of the overflow of the water and slime, this being invariably at the side of the vat, taking the form either of a discharge all round the rim or through two or three slat gates. With these rotating distributors, as usually arranged, there is much commotion at the side of the vat, sufficient, in fact, to keep sand in suspension and carry it out of the vat, this action being assisted by the slight throw to the circumference caused by the rotation of the water. The centre of the vat, where there is a tranquil zone, is the natural point for the overflow if it is desired to retain a maximum percentage of the sand, and the author has recently devised and protected a centre overflow gate for this purpose, which, used in conjunction with the automatic rotary distributor, will give a high percentage of sand attended by the advantages of automatic distribution. The utility of any device to keep in the sand becomes obvious when it is remembered that, apart from the economic considerations which make it desirable to catch as much of the sand as possible in the intermediate vats consistent with satisfactory filtration and a good extraction, the presence of sand in the slime is prejudicial to the treatment of the latter as at present carried out. Thus, with the rim overflow baffle boards are used, and with slat discharge gates the precaution is now taken of having a cloth screen in front of them to prevent the escape of sand through the joints.

City and Suburban Gold Mining Company, Limited.

Interior of Extractor House, showing Coir Filter Vats, Zinc Boxes and Sumps.

With few exceptions the contents of both intermediate and leaching vats are discharged through doors in the bottom of the vat, the number of these doors varying from one to eight per vat. The intermediate vats of the Treasury and the brick vats of the Crown Reef have side discharge doors, the filter bottoms of the latter plant being level with the ground, whilst the leaching vats of the New Primrose are discharged over the side. Of the numerous bottom discharge doors existent, that of McBride and Brown displays the most novelty, the door when released rests on two angle irons upon which it can be moved backwards and forwards by means of a lever; if no truck is beneath the door is pushed below the aperture and the boys throw the sand upon it, the accumulation being precipitated into a truck when the door is moved back. The large excavated leaching vats of the Langlaagte Estate are discharged by means of six travelling steam cranes which lower the bodies of the trucks into the vats and, after they have been filled by Kaffirs, raise and replace them on their saddles.

Of extractor houses there are many fine examples, but the description given of those of the Glencairn, the Simmer and Jack, and the Jumpers Deep convey an excellent idea of the latest practice, the first plant having zinc precipitation, the second one precipitating by electricity, and the third having both processes—zinc for its sand and electricity for the slime. The author, in his "Zinc *versus* Electricity in the Cyanide Process," entered fully into the merits of these precipitating agents, and here it need only be said, touching the future of the rival processes on these fields, that for sand zinc is likely to hold its own against electricity, but for dealing with the extremely dilute solutions used in the present process of slime treatment electricity is, as yet, without a rival, but a new method—a combination of zinc and electricity—is about to be tried.

Taking, first, the arrangement and plant of zinc extractor houses, we have the leaching pipes—one from each vat—discharging into coir filter vats, usually three in number, where the solutions are cleared of suspended matter and then passed to the extractor boxes, the number and dimensions of which vary with the size of the plant. The capacity of the extractor boxes is in many cases too small to obtain the best results; a good basis for designing a box is to allow a cubic foot of space for zinc shavings for every $\frac{1}{2}$ ton of solution to be passed through the box per twenty-four hours. With reference to the clean-up—the collection of the gold slime and its conversion into bullion—many cyanide managers have their own particular method, but in most cases the practice consists of collecting and filtering the gold slime with subsequent drying and roasting, after which it is mixed with fluxes and smelted. A better practice is to collect the gold slime and pass it through a filter press, badly corroded and crumbled zinc being also included in the clean-up receiving an acid treatment previous to going through the press. Mr. Blane's arrangement at the Glencairn is a convenient one for this work. The use of the filter press in connection with the clean-up is to be recommended; it saves much time and puts the slime into a very convenient form for handling. After leaving the filter press the slime is dried and is ready for fluxing. Of the two types of smelting furnaces used—the single-box type and the reverberatory—both have strong adherents. The reverberatory furnaces are slow to heat but get through the work quickly and are found more comfortable to manipulate.

For producing the zinc shavings required for the precipitation Liddell's machine lathe is finding many adopters and when properly adjusted and attended to it has a great capacity and gives little trouble. In connection with this machine it is worth noting that file-sharpened cutters of untempered steel have proved most satisfactory.

For dealing with the water and solution centrifugal pumps are used with few exceptions, and for generating a vacuum geared horizontal and vertical pumps and also ejectors are at work.

With electrical precipitation the boxes which are used are of greater size than zinc boxes. The recent alterations which have been effected in them consists of having the lead cathode in the form of strips and shavings, in simplifying the connection with the electrodes, mercury wells being now used, and in the boxes having fewer compartments. Plants with electrical precipitation—Siemens-Halske process—are now provided with a vat into which the prussian blue, formed during the precipitation, is periodically discharged, and to melt the auriferous lead for casting into ingots they have a reverberatory furnace, but the cupellation of these ingots is, with one exception, the Simmer and Jack, carried out at a Customs works.

For sumps we have round vats and oval and rectangular excavated ones, the capacity and number varying greatly, three vats being ordinary practice. In laying out plants every endeavour is made to place the sump vats at such a level that the solutions gravitate to them from the boxes; in many cases this necessitates placing the sumps in an excavation. Strong solution sumps have provision for mixing their liquor; in many cases it takes the shape of a return pipe from the pumps which enables the latter to circulate the solution. The arrangement recommended by the author is a rotating air or steam agitator such as he has already described; by its means the contents of a large vat can be thoroughly mixed in a few seconds whilst to effect a satisfactory mixing by pumping occupies much time.

In the latest plants the bore of the leaching pipes has wisely been reduced compared with early practice, and pipes of $1\frac{1}{2}$ in. diameter are now frequently fitted to vats of even 300 tons capacity. This reduction in size prevents, to a great extent, the practice of sending the

Geldenhuis Estate Gold Mining Company.

Cyanide Works.

solutions through the extractor boxes at an undue rate, a practice which is much too common.

Concerning the present treatment of the slimed portion of the mill pulp the process is too new and the plants too few for much to be said thereon, but regarding the process itself there is every indication that it has come to stay. Its development has occasioned costly experiment and taken much time, but the resultant process is of a comparatively simple nature, and practical work has put beyond doubt the feasibility of applying it with commercial success on these fields to slime of 3 dwt. value under favourable conditions. It is still in its infancy, however, and will undoubtedly be further improved, but even now results are being obtained which promise little, if any, scope for prospective rival processes. Of the plants at work the general design of that of the Crown Reef with its excavated settling pits and pump agitation has not met with the same approbation as the Robinson type with paddle agitation although the former plant is giving good results, but the works to be erected at the Jumpers Deep surpass all existing ones in design and will doubtless give an excellent account of themselves when in operation.

The metallurgical practice of the Rand is almost exclusively one of wet milling with amalgamation and cyanidation, and there is nothing pertaining to the very limited practices of chlorination of concentrates, and the equally limited dry crushing of ore with direct cyanidation, that calls for special comment.

The above lines give but a glimpse of present day metallurgical engineering on the Rand, for the subject is far too wide to be embraced with anything like completeness by these few pages. It was not purposed to touch upon the routine of the mills and cyanide works in them, but where this has been required for elucidation it has been done, and now, having portrayed to some extent the plants and practice into which the experience and ideas of Rand engineers have resolved themselves, it is fitting and of interest to conclude by citing some of the financial results achieved, and I cannot do better than quote some of the figures embodied in a statement placed before the recent Mining Commission in Johannesburg by Mr. Hennen Jennings, consulting engineer to Messrs. Eckstein & Co. In this Mr. Jennings shows that owing mainly to improvement in method, plant, and process, the working costs of the Robinson Mine, which were over 72s. per ton in 1888, had been reduced to barely 31s. per ton in 1896, although the latter figure included two more metallurgical processes ; that at the New Heriot the total working costs in 1894 were 32s. 11d. per ton, whilst in 1896 they were 26s. 10d. per ton, with an increase in the extraction which raised it to 85·3 per cent. ; and that the milling costs at the Crown Reef which in 1890 were 11s. per ton, had been brought down to 3s. per ton in 1896, whilst the original extraction of 57 per cent. had been gradually increased until it had attained a total of 86 per cent., this being the extraction for the half-year ending March 31, 1897 when the slime works was added. Mr. Jennings also pointed out that the class of machinery on these fields was the most perfect of any gold field in the world, and that the average cost of milling of twenty-nine companies was 3s. 8d. per ton, a figure which compared very favourably with milling costs on other fields where the working was under similar conditions.

Regarding cyanide practice the Rand indisputably leads in the matter of low costs, for it is not unusual for some of its plants to work for 3s. per ton, and at one mine—the Geldenhuis Deep—the process has been run at a cost of a fraction over 2s. per ton treated.

APPENDIX.

THE

ECONOMICS OF RAND MINING.

APPENDIX.

THE ECONOMICS OF RAND MINING.

Mr. HENNEN JENNINGS' STATEMENT

Read before the Mines Commission, Johannesburg, 12th May, 1897.

Mr. HENNEN JENNINGS, Consulting Engineer to Messrs. H. Eckstein and Co., was then called to give evidence.

Witness said : I have resided in Johannesburg since the latter part of 1889.

The Chairman : On what points do you wish to give evidence?

Witness : I wish to give evidence in accordance with the Chairman's statement, in order to bring to light the actual state of the mining industry of the Witwatersrand Goldfields, and the reasons for the same, and my candid opinion on the present condition of affairs. I have a statement to make which I wish to read. Witness then read the following statement :

I wish to make a statement more or less summing up, including and extending the evidence or preceding witnesses, and to vitalise the facts and statistics set before you by a connected linking of these facts in a logical and orderly manner, and to give you my candid ideas as to the actual state of the mining industry.

The magnitude and wealth of these Goldfields have been examined, discussed, and written upon by very many able men from all parts of the world, and they have all agreed in stating that this deposit is unique in its characteristics, and contains vast possibilities.

These reefs, which are conglomerate beds, have been traced for some 50 miles, showing varying thickness and gold value, and in one point proved by the Bezuidenville bore hole to a vertical depth of 3,130 feet

Mr. Hamilton Smith has happily described the fields by stating that the excellence of these mines is not due to their exceeding richness, but rather to the large continuous bodies of ore of a moderate grade, and has recognised the necessity for the best possible mechanical plants, and the most skilful and economical management. He, moreover, was of the belief that if the management were radically bad, not more than three or four mines could have yielded considerable profit. He also estimated (January, 1893) that within the then recognised paying area of the Witwatersrand goldfields, down to a vertical depth of 3,000 feet, there was probably 325 million pounds sterling worth of gold to be extracted.

His statement has been corroborated by Mr. Schmeisser, a German Government engineer, who estimated at the end of 1893 that down to a vertical depth of 3,900 feet, and for a lateral extent of eleven miles, there were possibilities of £349,367,000.

Professor Becker, of the United States Geological Survey Department, who visited these fields in 1896, estimates the possibilities within twenty miles of Johannesburg, to a vertical depth of 5,000 feet, at about 700 million pounds sterling.

Hitherto, in all gold-mining regions of the world, gold mines have been considered highly speculative ventures, and liable at any time to give out; they have nowhere else such advantageous natural conditions as here for making a staple permanent industry, nor the same justification for the great expenditure of initial capital on their equipment and development. Gold is here supplemented by coal in close proximity, ample water supply and favourable climatic conditions. Of course there are fluctuations even here in the richness of the different mines, and there are unaccountably rich, medium, and poor mines in juxtaposition or in distinct sections.

There is no doubt there are sections of the Rand which can be continued to be worked at a good profit at the present rate of costs ruling on these fields, but there is a far larger amount of ground that will not be worked, or only tried and then abandoned, if the conditions prevailing as to costs are not lowered, and the predictions of the eminent authorities I have quoted can only be realised by all parties interested doing everything in their power to obtain highly efficient working at low cost, and thus increasing the scope and possibilities of the whole fields.

Take Professor Becker's estimate of 700 millions sterling. At the present rate at which we are now taking out gold, these fields would have a life of about ninety years, but the rate at which we extract this gold will be constantly increasing if the working conditions are rendered more favourable, and the life of the fields will become correspondingly shorter; this takes into no account the working of our southern low grade reefs, but only the Main Reef series. Professor Becker's estimate, too, as well as Mr. Hamilton Smith's and Mr. Schmeisser's, do not include the Klerksdorp, Potchefstroom, Heidelberg, Lydenburg, Barberton, nor other districts, which, according to the figures for 1896, contributed approximately 9 per cent. to the total gold production of the South African Republic.

The State Mining Engineer, in the course of the proceedings of this Commission, has stated that there have been 185 gold mining companies floated in this Republic, with a nominal capital of £54,000,000.

On the Witwatersrand fields there have been about 5,500 stamps erected. The annual report of the Chamber of Mines shows that on an average 3,470 stamps were running during 1896; consequently, it would appear as if 2,030 stamps had been stopped; but this is not really so, as many have been dismantled and new ones have replaced them. But it is a fact that there are on the fields many companies with a large number of stamps that have suspended operations, and there are several others which, during this year, will probably follow suit.

The Chamber of Mines report shows that in 1896 there were fifty-six companies in operation; while the statements from the Chamber of Mines and the Association of Mines for March show that there are now only forty-seven companies, with 3,275 stamps working.

On the other hand, great energy is being shown by Deep Level and other companies in pushing forward work with the object of starting more stamps, and it is estimated that about 1,000 new stamps will start during the current year, if conditions are favourable.

I have gone to considerable pains to obtain as far as possible the last annual report of all principal mining companies working in 1896, and which have been working continuously during the periods covered by their last annual report, and to analyse these reports—the gold returns being according to the sworn statements of their managers, and the accounts being in each case signed by the auditors and secretaries. It should be noticed that, in dealing with these twenty-nine companies, the period covered is not necessarily the year 1896 only, but embraces the actual period covered by each individual report, and often this is for part of 1895 and part of 1896, and in several cases includes periods longer than one year.

This table I consider most remarkable, in that it is compiled by a private individual from published statements given freely to the world, and anybody can make out the same table who obtains the same reports. In this connection I would state that in discussing the matter of statistical information about these fields with different eminent mining men from all parts of the world, the consensus of opinion is that more generous and accurate information is given on these fields of the working of the mines than is the case in any other part of

		No. of Stamps	Tons Mined	Tons Sorted	Per Cent. Sorted	Tons Milled	MILL RETURNS			SECONDARY SANDS AND (	
							Value	Yield per Ton		Tons Treated	Value Yield
TOTALS:							£	s.	d.		£
A. 18 Companies		1,625	2,659,209	169,307	6·367	2,489,902	3,721,811	29	10·74	1,695,652	1,566,176
B. 7 do.		561	794,779	128,470	16·164	666,309	854,689	25	7·85	449,935	366,608
C. 4 do.		270	288,271	...	...	288,271	286,945	19	10·89	280,572	150,254
29 Companies		2,456	3,742,259	297,777	7·95	3,444,482	4,863,445	28	2·86	2,426,159	2,083,038

NOTE.—The dividend paid by one of the companies in the group

TREATMENT: CONCENTRATES				TOTAL RECOVERY.			WORKING EXPENSES PER TON									
VALUE YIELD PER TON						Per Ton Milled		Mining		Mine Development and Redemption		Total Mining		Transport		Milling
Treated		Milled		£												
s.	d.	s.	d.	£	s.	d.	s.	d.	s.	d.	s.	d.	s.	d.	s.	d.
18	5·68	12	6·96	5,287,987	42	5·70	12	10·37	3	10·04	16	8·41	0	1·51	3	7·16
16	3·55	11	0·05	1,221,297	36	7·90	14	4·44	4	5·58	18	10·02	0	1·63	3	8·28
10	8·53	10	5·09	437,199	30	3·98	15	0·28	7	5·80	22	6·08	0	1·85	4	3·70
17	2·05	12	1·14	6,946,483	40	4	13	4	4	3·12	17	7·12	0	1·56	3	8·08

showing a loss during its financial year is explained by the fact that although it was declared in January, 1896, it referred partly to a

AS SHOWN BY THEIR LAST PUBLISHED ANNUAL REPORTS.

ON BASIS OF TONS MILLED					GOLD PRODUCED FROM ALL SOURCES		Total Working Expenditure, including Depreciation, but excluding Capital Expenditure and Permanent Works	PROFIT	
Secondary Treatment	General Charges not segregated	Total Cost per Ton	Depreciation	Total Cost with Depreciation		Average Value on milled tonnage basis per ton		Total	Per Ton
s. d.	s. d.	s. d.	s. d.	s. d.	£	s. d.	£	£	s. d.
3 1·33	2 6·75	26 1·16	3 6·19	29 7·35	5,346,363	42 11·33	3,686,710	1,659,653	13 3·98
2 9·94	2 6·64	28 0·51	4 10·31	32 10·82	1,222,369	36 8·29	1,096,151	126,218	3 9·47
4 9·40	4 7·45	36 4·48	2 7·83	39 0·31	439,533	30 5·93	562,502	...	...
3 2·34	2 8·90	27 4	3 8·44	31 0·44	7,008,265	40 8·31	5,345,363	1,662,902	9 7·87

preceding period.   This company has also written off heavily for depreciation of old plant and mine development, and has thus considerably increased its

Loss		Tonnage of Ore Developed and now in Reserve	Dividends Declared and Paid during Period	Nominal Capital	Issued Capital	Value of Issued Capital at Market Rate 24th April, 1897	DIVIDENDS PAID DURING PERIOD	
Total	Per Ton						Per Cent. on Issued Capital	Per Cent. on Issued Capital at Market Rate 24th April, 1897
£	s. d.		£	£	£	£		
...	...	4,390,426	1,729,103	6,981,000	6,881,586	15,123,876	25·127	11·433
...	...	2,543,243	...	6,945,000	6,945,000	14,010,937	...	...
122,969	8 6·38	522,812	32,000	755,000	755,000	531,897	4·450	6·016
...	...	7,456,481	1,761,103	14,681,000	14,148,221	29,666,710	12·446	5·970

working expenditure on the basis of my analysis.

the world. In addition to the companies' reports, we have also the vast amount of accurate information collected by the State Mining Engineer, the Chamber of Mines, and the individual enterprise of Mr. C. S. Goldmann, so that the investing public should certainly be cognisant of all vital facts here with such statistics before them ; and if they exaggerate the possibilities of the mines, we have nothing to reproach ourselves with.

At the same time it is regrettable that more publicity has not been given to the work of the State Mining Engineer's Department as regards the statistics of the gold industry. The Government should have been proud of this collection, and had it published in all languages and distributed to the world. They have, I think, been negligent in this respect, for, although their information is most valuable, it is inaccessible to the majority of investors in these fields.

Now, to return to my own statistics. The companies included in my list are as follows :—City and Suburban, Crown Reef, Durban Roodepoort, Ferreira, Geldenhuis Deep, Geldenhuis Estate, Ginsberg, Glencairn, Henry Nourse, Johannesburg Pioneer, Jubilee, Jumpers, May Consolidated, New Comet, New Heriot, New Primrose, Robinson, Salisbury, Simmer and Jack, Wolhuter, Worcester, George Goch, Langlaagte Estate, Langlaagte Royal, Meyer and Charlton, New Midas Estate, Roodepoort United, Van Ryn Estate, Wemmer.

It will be noticed that there are twenty companies omitted from the total now working ; a large proportion of the reports of these do not cover a full year's work, while of others I was unable to obtain copies.

The only dividend companies of 1896 omitted from my list are :

Stanhope	. £1,700
Langlaagte Block B (preference shares)	6,500

The above twenty-nine companies I now divide into three groups, summing up the yields, working costs, etc., under the heads of mining, milling, secondary treatment, etc., and also give the dividends paid during the period, the capitalisations, and the interest per cent. on such capitalisations.

These three groups are :

A. Mines that have paid dividends during the period covered by their last annual report.

B. Mines that have shown a profit, but for various causes have paid no dividends during the period of their last annual report.

C. Mines that have worked at a loss during the period of their last annual report.

The summary of this statement is as follows :—

This statement shows that even at present depressed prices the public places a value on the shares of the twenty-nine companies which is more than double the amount at which they were originally capitalised, and that, taking the dividend actually paid, the interest obtained by investors in the eighteen dividend-paying companies alone is 12·4 per cent., and that if these dividends be distributed over the whole twenty-nine companies in my list, the interest obtained is only about 6 per cent. What must it be for the whole fields ? It appears to me that too much stress has been laid by this Commission on company capitalisation. What, after all, does it mean to the intrinsic investor what the capital of a company is, so long as he can buy into it at what he considers a profitable price ? Are not the original capitals mere guesses to establish rates of division of interest, and if the guess is too small, may not as much harm be done as if the guess is too high ? The conscientious engineer and examiner of a mine simply regards capitalisations as counters on which he places value in accordance with the probabilities of its earning power.

From this statement of yields and costs it is seen that the cost per ton of ore milled is shown in two ways :

(a) Cost per ton, with depreciation of plant.

(b) Cost per ton, without depreciation.

Depreciation is considered by many merely a book entry, and unfair to include in legitimate working costs.

On the other hand, most companies do not include permanent main mine works in their running costs or current construction work ; therefore I think the mean of these two costs a fair average for the fields, before any dividends are distributable, and we thus obtain for the total working costs of the fields, as shown by the twenty-nine companies I have mentioned, 29s. 4d. per ton, and the total yield from all sources, including mill, cyanide works, etc., 40s. 8·31d. per ton.

The cost thus arrived at is seen to correspond with all reasonable degree of accuracy with the statement made by the Chairman at the Rand Mines meeting, and I therefore beg to hand in as evidence that portion of his speech which deals with this subject.

It will be noticed that there are seven million tons of ore developed by these twenty-nine mines, which is equivalent to a cash asset of at least £1,750,000.

Another interesting feature to be noticed from the tables is that the yield from the secondary treatment is shown to be 12s. 1·14d. per ton on the basis of the tonnage milled, and working costs 3s. 2·34d., the profit from this treatment therefore figuring at 8s. 10·8d. From this it is clearly evident that of the total profit of 9s. 7·87d. obtained by the combined treatments, no less than 8s. 10·8d., or 92 per cent., came from the secondary treatment, without which obviously only an extremely small number of the very richest mines here could ever have paid dividends.

This is a strong illustration of what intelligent metallurgical and engineering skill has done for the prosperity of these fields.

I also beg to submit a statement showing the analysis of the working expenditure of six prominent companies for the year 1896. In this sheet are given the details of costs under labour and supplies of the following :— Crown Reef, Henry Nourse, City and Suburban, Robinson, New Heriot, Geldenhuis Deep.

The summary of this statement is as follows :

	Costs per ton.		Per cent. of total cost.
	s.	d.	
Native labour .	6	9·62	23·73
Native food .	1	2·24	4·14
White labour, salaries, etc.	8	7·78	30·18
Coal . . .	2	4·35	8·24
Dynamite .	2	10·13	9·92
Fuse and detonators	0	1·97	0·57
Cyanide .	0	8·12	2·36
Zinc .	0	0·62	0·18
Mining timber . . .	0	2·98	0·87
Timber, deals, etc.	0	4·05	1·18
Steel .	0	4·21	1·22
Oil, grease, and paraffin . .	0	3·73	1·09
Candles	0	4·15	1·21
Ropes, steel and manilla	0	0·93	0·27
Electric spares . . .	0	1·47	0·43
Mill spares, shoes, dies, cams, cam shafts, stems, mortar boxes, screening, etc.	0	4·59	1·34
Trucks, wheels, and rails . .	0	2·24	0·65
Sundry stores : Bar iron, bolts and nuts, buildings, machinery, assay chemicals, pipes and pipe fittings, etc.	2	0·73	7·19
General charges : Insurances, licences and rent, printing and advertising, sundries	1	5·99	5·23
TOTALS	28	7·90	100·00

This again corresponds so closely with the statements made by the Chairman at the recent Rand Mines meeting that I beg again to put this portion of his speech in evidence. It will be noticed that on the sheet I submit each Company's Secretary has placed his signature as a voucher for the accuracy of the statements.

In support of these figures as being a criterion for the whole Rand, I beg to state that the following statistics were compiled from the State Mining Engineer's report for 1895 :

	£	Per cent. of total cost.
White labour	2,400,000	34·3
Kaffir labour	2,000,000	28·6
Coal	700,000	10·0
Explosives, i.e., dynamite and gelatine	600,000	8·6
Mining timber and sawn lumber	300,000	4·3
Cyanide	240,000	3·4
Meat, mealies and mealie meal (being for the most part food for Kaffirs)	250,000	3·6
Iron	85,000	1·2
Candles and paraffin	95,000	1·4
Tools	70,000	1·0
Steel	65,000	0·9
Mercury, zinc, and other chemicals	45,000	0·6
Caps, safety fuse, rope, cement, etc.	150,000	2·1
	£7,000,000	100·0

This is seen not to correspond exactly with the statement I have given; a simple explanation of which is that the State statistics are for a different period and cover an expenditure of £7,000,000, which includes the non-producing as well as the producing mines, while the figures in my statement only cover an expenditure of £1,300,524.

I again wish to put in evidence the State Mining Engineer's figures regarding wages paid on these fields to white labour, taken from the report of 1895, which shows :

Occupation.	Number.	Average monthly wage. £
Shift bosses	185	33
Miners	1,430	23
Rock-drill machine-men	956	17
Trammers	226	18
Engine-drivers	765	24
Pump men	129	23
Stokers	89	19
Carpenters	1,058	26
Smiths	638	26
Mechanics and fitters	900	26
Bricklayers	75	22
Stonemasons	213	29
Daily labourers	149	18
Mine and store clerks	287	23
Amalgamators	291	23
Cyaniders	217	22
Concentrators	35	22
Vannermen	32	20
Smelters	21	26
Various workmen	472	21
	8,168	£461

making an average of £23 7s. 10d. per man per month.

This amount is seen to be somewhat lower than that given by the Manager of the Crown Reef in his evidence, which figured out at £24 1s. 10d., and which is, I believe, representative of the five other companies dealt with in this statement, but I would prefer to deal with this subject from the State statistics basis, and if there are any errors in the statement as regards the State Mining Engineer's Department, I beg that he will correct them so that they can be put in this statement on their right basis.

I also wish to vitalize all these statements by more or less culling from the history of the six companies I have given, and whose workings I am in a position to state with accuracy, and to show how these yields and costs have been arrived at, and how these whole fields compare with some of the yields and costs of gold mining in other parts of the world.

We will take the Robinson Company as the typical rich mine, and follow a little of its history. It will be noticed from its published annual reports that it first figures as a gold producer in the year 1888, and that it commenced with ten stamps, which ran up to 1889 before being superseded by forty new ones; that the yield per ton for its first year's work was 272s. 7·04d., and that the working cost, which then only included mining and milling, was 72s. 1·04d per ton ; that the extraction was 65 per cent. ; the machinery then erected was of a crude nature, and the mine workings rather of a prospective than a permanent kind.

Go to any old prospector and ask him the method in which he looks for gold in a new region. He will tell you that he has no great geological knowledge, but that he knows gold when he sees it ; that he goes over the surface and wherever he finds an outcrop he tries it with the pan. He goes over the whole extent of outcrop that is open to him and naturally picks out the richest points at which to start work. He commences on a small scale and he works these rich parts, and as they give him profit so he extends his operations. So started the Robinson, and so was the Rand developed.

The richest mines are started first, and the richest places in these mines are selected for a start. It would have been folly to have done anything else. In the early stages there was no need of a highly trained engineer. In fact, a conscientious one would have told his principals that he had never seen any deposit like this, and he would have to be guided by actual disclosures in order to intelligently advise them.

With this digression let us return to the history of the Robinson Company.

ROBINSON G. M. CO., LTD.

The nominal capital of the Robinson was first £50,000 in £1 shares, of which £5,000 were working capital. This small original working capital was due to the insistence of the original owner of the ground.

The original owner was bought out after a time, and on January 24, 1889, the capital was increased from £50,000 to £53,375, in order to acquire six claims leased to third parties before formation of company. This again was changed on February 16, 1889, to a capitalisation of £2,750,000 in 550,000 £5 shares, to meet the public, who had placed a value of £60 or £70 upon the original £1 share.

All of this capital was issued to shareholders with the exception of 16,250 shares held in reserve, the area of the gold-bearing ground being equivalent to 106 claims, which has since been increased by purchase to 136 claims.

The financiers who controlled the workings of this mine realised the large capitalisation, and their endeavour has been to justify it by actual returns from the mine, and their success is demonstrated by the fact that even on the present low market value of to-day, viz., £6 14s. per share, the capitalisation of this Company is held by the shareholders at £3,685,000, and it is considered in Europe one of the Consols of the industry.

On this capitalisation the last year's profit shows 10·21 per cent.

Now, how was this brought about ? The ten-stamp mill was replaced in 1889 by forty stamps, the forty were extended in 1891 to sixty, the sixty extended in 1894 to seventy, and in 1895 the mill was further extended to 120 stamps.

The yield and working costs during these periods were as follows :

	Working cost per ton, including depreciation.		Yield per ton.	
	s.	d.	s.	d.
1888	72	1·042	272	7·04
1889	65	11·846	182	7·24
1890	65	1·865	113	2·94
1891	52	5·575	103	5·53
1892	46	5·997	95	6·78
1893	42	1·097	101	0·02
1894	41	4·736	97	4·94
1895	30	0·913	80	5·67
1896	30	11·096	69	10·20

Mining was carried on in three reefs in this mine, known as the South Reef, Main Reef Leader, and the Middle Reef; the latter was afterwards discovered to be simply an overlap of the South Reef, and disappeared altogether at the third level.

The ore in the upper level down to a depth of about 210 ft. was what is known as "free milling." The dip was forty-two degrees, but has gradually flattened until at the greatest depth, viz., 1,484 ft. down on the incline, it is only twenty-nine degrees. At a depth of 210 ft. the matrix of the conglomerate pebble formation changed from an oxidised to a pyritic character, and the mining became somewhat more difficult and costly.

The difficulties in obtaining satisfactory returns by simple plate amalgamation then became greater, and one of the problems facing this Company was to get an adequate return of gold from the ore, as tailings leaving the mill averaged as high as 14 dwts. per ton.

This Company was the first on the Rand to successfully run frue vanners. It supplemented this work by the erection of a chlorination plant, which not only dealt with its own concentrates but also those of other companies, and this chlorination plant has produced gold to the value of over £860,000 since it started.

The Company was also the first to introduce on a large working scale the cyanide process. It replaced the first original works by larger ones, and has expended in connection with this branch of the industry over £40,000.

The treatment of slimes was also introduced on a large scale at this mine, and in connection with the Rand Central Ore Reduction Company it has expended £60,000 to £70,000 in this direction.

The total amount of money put into buildings, plant, improved processes, etc., by this Company has amounted to . . £426,736

Development 355,528

Making a total outlay of £782,264

and the reserve ore in sight in the mine is 441,506 tons.

Granted that a large proportion of this money was obtained from the ground, it was put back into the ground, and it aided all other companies in improving their appliances, and in their own case has brought up the gold extraction from 65 per cent. in rich free milling rock to over 90 per cent. in the pyritic ore worked at the present day, and has thereby given encouragement and impetus to all mining on these fields.

In dealing with the metallurgy of the ores on the Rand, the ordinary process of milling, concentration and chlorination, though adopted at this mine, did not prove as successful as in other parts of the world, owing to the way in which the gold was found in the rock, i.e., in very minute particles; and this great percentage has been obtained by the initiation of a comparatively new system in metallurgy which has had its growth and development on these fields, viz., the treatment of tailings and slimes by the cyanide process, and the Robinson Company was one of the early pioneers in this direction.

It will be noticed that a remarkable decrease in costs has taken place from the early stages of the mine to the present time.

P

Milling and tramming costs have been reduced from 18*s.* 8*d.* to 3*s.* 8*d.* per ton; mining costs 36*s.* 3*d.* to 17*s.*, inclusive of development.

Cyanide costs started at 13*s.* 6*d.* per ton treated, including royalty, and were reduced to 3*s.* per ton treated last year. The total cost per ton on a milled basis was 72*s.* 1*d.* in 1888, and is now 30*s.* 11*d.*, including two more metallurgical processes, working in harder ground, and all current capital expenditure as well as depreciation, showing a reduction in cost of 41*s.* 2*d.* per ton.

In comparing the cost of the Robinson with other mines it must also be remembered that, although little sorting is done on the surface, a considerable quantity of waste is eliminated below.

These are indeed startling reductions, and, from an engineer's point of view, I cannot see how the efficiency of this Company's works can be greatly increased.

The Board have used every endeavour to obtain men of acknowledged excellency in their departments from all over the world.

The manager of the Alaska Treadwell, the working costs of which mine have so prominently figured before the public, was induced to take the management of this Company in 1892, and, though not alone, a great deal of the credit of the reduction in costs is due to his instrumentality and to his wonderful power of dealing with the employés of the Company. His motto among his officers was to impress them with the idea that in each department they were to work as if the Company were their own; and with great firmness he combined great kindness, and was loved as well as respected.

The work done at the Robinson has had its effect on other mines, and without a Robinson Mine you would not have a Heriot, and without a Heriot you would not have a Geldenhuis Deep.

NEW HERIOT G. M. CO.

The Heriot Company is an illustration of a not exceptionally high grade mine starting with inadequate working capital, poor and meagre equipment, and being obliged to suspend operations owing to the fact that it could not be made to pay with the appliances and funds at the disposal of the Company.

The Company was formed in August, 1887, with a capital of £50,000 in £1 shares, the vendors receiving £4,000 as the price of the property, which consisted of twenty-nine claims.

The battery commenced work in January, 1888; the initial yield after six months was valued at £2 16*s.* per ton, and the actual expenditure £1 19*s.* 8*d.*

The capital was increased in October, 1888, by issuing to shareholders 10,000 shares at 30*s.*, and three dividends of 5 per cent. were declared, but the Company again ran out of funds.

In December, 1889, the Board was authorised to increase the capital to £75,000 by creating 15,000 new shares. But no tenders for these shares were received, and it was not until April, 1890, that they could sell even at par the small amount of 1,000 shares, and these only on the condition that they were to be redeemable at 30*s.* within six months. In the following month 5,000 were allotted to an applicant for £7,500.

Attempts were made to obtain loans, but these were fruitless owing to the restricted borrowing powers of the Directors, which were limited by the trust deed to £3,000; and, moreover, the bank not only refused to sanction an overdraft beyond that amount, but expressed disapproval of the debt already incurred.

During the year ending July, 1890, the mill of twenty-five stamps crushed 8,873 tons, yielding gold to the value of £3 1*s.* 4*d.* per ton, the working costs being 52*s.* 6*d.* for mining and milling only.

During 1891 the mill practically stopped work. Early in 1892 the New Heriot Company was formed, the capital increased to £85,000, and the management and finances put into strong hands.

The mine was developed and thoroughly equipped with the cardinal idea of a centralisation of power. The cyanide process was introduced, and the new works were started towards the end of 1893. The total working costs in 1894 were 32*s.* 11*d.*, in 1895 27*s.* 4*d.*, in 1896 26*s.* 10*d.*, and the extraction brought up to a total of 85·3 per cent.

This is now considered one of the model mines of the Rand, and the manager is a practical man who has grown up with the industry here.

The capital was increased in 1895 to £115,000, in order to acquire more ground, and the total number of reef-bearing claims is now fifty.

HENRY NOURSE G. M. CO., LTD.

This Company was floated at Pretoria in April, 1887, with a capital of £35,000, in £1 shares, of which 24,000 shares were given for the property, and 11,000 shares were issued against £11,426 5s., which constituted the original working capital. This sum was soon exhausted, and a special general meeting of shareholders was held in Pretoria on June 4, 1888, when the capital was increased by 15,000 shares, which the Directors were instructed to issue at not less than 30s. each. At the first annual meeting of shareholders held on August 14, 1888, the Directors reported that owing to this limitation they had been unable to dispose of these new issue shares, excepting 2,670 shares taken by Sandycroft's agent to settle an amount due to them and to pay for a new fifteen-stamp mill which had been ordered. The balance-sheet to June 30, 1888, shows that 1,642 ozs. 9 dwts. of gold had been won, so that the original fifteen-stamp battery must have started about March of that year, but no record of the number of tons crushed appears to have been kept. On June 30, 1888, the Company's cash was exhausted, but there were on hand 12,330 out of the 15,000 reserve shares created in June, 1888.

At December 31, 1888, the west shaft had been sunk to a depth of 103 ft., the central to 16 ft., and the eastern to 39 ft. The report to that date states that 2,330 of the reserve shares had been disposed of at 31s. per share, and the balance of 10,000 at 56s., thus placing the Company in funds ; and that an order for a fifteen-stamp battery had been increased by forty stamps, making fifty-five stamps in all, to make with the fifteen-stamp mill then running a battery of seventy heads. This battery was of too light a pattern, and was never erected, but was disposed of as an opportunity occurred.

At December 31, 1889, the Company's indebtedness had increased to £24,905 17s. 9d., and at June 30, 1890, to £26,946 3s., and at a special general meeting of shareholders held on September 10, 1890, the capital, then £50,000, was increased by the creation of 50,000 shares to £100,000. Of these 50,000 shares 37,500 were offered to shareholders at 30s. but were not applied for. Eventually these shares were taken up and came into strong hands.

The report of December 31, 1891, shows that it had been necessary to pledge the Company's assets as security for an overdraft at the Standard Bank, and the above arrangement as to the issue of the reserve shares was the best the Directors were then able to make.

At June 30, 1892, the Company had paid off its liabilities and had a cash balance of £25,000. By December 31, 1892, the fifteen-stamp battery hitherto running was increased by twenty stamps, and on December 12, 1892, a cyanide plant was erected near the old battery and commenced work.

During the next half-year five more stamps were added to the battery, and extensive additions were made to the pumping and hauling equipment, and at June 30, 1893, the Company was again in debt about £17,000.

The mine was, however, opening up well, and showing such good returns that a special meeting held on March 7, 1894, sanctioned the increase of the capital to £125,000 by the creation of 25,000 shares, which were taken up by shareholders and guarantors at 40s. each. The capital thus raised and profits accruing from mining operations were expended in the erection of a first-class plant on the basis of sixty stamps, the total expense incurred in the equipment and buildings amounting in all to £200,000.

The total number of claims in this property is about eighty.

The debt that was incurred in connection with the equipment was not wiped off until the middle of 1896, and the first dividend was 30 per cent., declared in December, 1896, and paid in February, 1897.

The yield from this Company has been fairly uniform, and shows an average extraction from the start to the present day of about 82 per cent.

The costs have been reduced from their maximum of 56s 3d., to the minimum, 35s. 7d., shown in my table of working costs.

Even the present costs are high, owing to the small lateral extent of the property, which has necessitated the working through three shafts, and also to the fact that for the most part only one narrow reef has been worked, and 25 per cent. of the rock hoisted from the mine sorted out as waste; the rock has also been exceptionally hard. The small dip area for the major extent of the mine is due to the fact that, on starting, the conglomerate beds showed a declination of about 80 degrees, which would allow mining to be conducted at considerable depth before reaching the southern boundary.

The beds, however, within a horizontal distance of 240 ft. have gradually changed their dip to forty-six and fifty degrees, thus permitting the deep level company to commence mining operations at a comparatively shallow depth.

CITY AND SUBURBAN G. M. CO.

This Company was formed in 1887 with a capital of £50,000 in £1 shares, of which the vendors received £30,000, and then certain other interests were acquired, reducing the working capital from £20,000 to £8,900.

The capital was increased in

1888 by	1,000 shares, realising	. . .	£15,548
1889 by	5,000 „ „	. .	62,950
1892 by	10,000 , „	.	65,000
1893 by	10,000 , „	. . . . . .	86,250
So that the total working capital subscribed amounted to		. .	£238,648

It must also be added that in 1895 the capital, which then stood at £85,000, in £1 shares, was transformed into 340,000 shares of £4 each.

The mining area of the Company is 150 claims (about), of which about twenty claims have been worked out up to end of 1896, including poor ground and pillars. The total tonnage crushed up to this period is 670,463 tons, and the gold bullion recovered from same 428,794 ounces.

The total cash receipts from all sources have been	£1,838,484
And the total cash expenditure on property, development, equipment and working expenses to end of 1896	1,559,695
Leaving a profit of	£278,789

out of which £251,661 has been paid in dividends to shareholders, and the balance of £27,128 is represented by cash, stores on hand, etc.

The equipment of the mine, the cost of which stands at £524,110, includes the development of ore reserves amounting to 375,895 tons.

The first ten-stamp mill started crushing in August, 1887, to which was added another ten stamps in June, 1888. In May, 1891, a new thirty-stamp mill was started, to which was added the first twenty stamps two months later. This fifty-stamp mill worked on till November, 1895, when it was finally closed down, having been replaced by an improved plant of about eighty stamps started in July, 1894; forty stamps were added to the latter in July, 1895, and in September following a further forty stamps, making up 160 stamps at the new mill. The full plant, however, was not worked until July, 1896, owing to inadequate supplies of native labour.

Cyanide works were started in 1893 to treat the product of the fifty-stamp mill and the accumulation of tailings prior to that date. In July, 1894, a new cyanide direct-filling plant in connection with the new milling plant was started, which has since been extended for the treatment of coarse sands and concentrates, and double treatment has been adopted.

As no systematic samples of the ore were recorded prior to 1892, figures are not at hand to supply the percentage of recovery previous to that year. In 1892 the recovery by amalgamation only was 59·596 per cent.

of the value of the ore crushed, which has been raised to 81·95 per cent. for 1896 by mill and cyanide, showing that with improved methods the extraction has been increased by 22·36 per cent.

It will also be noted that the working expenses have been reduced from 60s. 4·56d. per ton in 1887 to 26s. 3·91d.· per ton in 1896, due to the improved method and appliances of mining and milling, and ore treatment, and working on a large scale, thus now making it possible to work profitably lower grade ore than formerly.

CROWN REEF.

Taking this Company next, which in early days was considered only a low grade mine, though having two very regular reefs running through it with an average stoping width of 4 ft. each.

Its history can be summarised as follows :

This Company was formed on April 1st, 1888, to acquire the lease of a mijnpacht on the farm Langlaagte from a private syndicate. It had an issued share capital of £100,000, in £1 shares, of which amount £14,000 was working capital. The share capital of the Company was increased to £106,000 in 1890 ; to £110,000 in 1892 ; and £120,000 in the latter half of 1892 ; the pront from the sale of shares being utilised to equip the property.

The milling power of the Company was thirty stamps at the start. This was increased by a new mill of forty stamps in 1890. To this mill a further twenty stamps were added in 1892. These two mills were replaced by a new mill of 120 stamps, with complete cyanide works in 1894, and slimes plant in 1896.

	£	s.	d.
Up to March 31st, 1895, the Company had expended on capital account, buildings, machinery, and plant, dams and reservoirs, etc. . .	308,963	0	9
Purchase of freehold rights	26,009	0	0
Sinking main shafts and driving main cross cuts	62,521	7	3
	397,484	8	0
Besides this, mine development charged to working expenditure cost	58,850	16	9
	£456,335	4	9

Since that time all expenditure has been charged against revenue account, and capital account has ceased entirely. During the two years ending 31st March, the Company has expended for :

	£	s.	d.
Sinking main shafts and driving cross-cuts . .	21,618	9	2
Mine development	64,674	10	10
Buildings and additions to plant	6,433	18	0
	92,726	18	0
Amount brought down from above	456,335	4	9
Making a total of	£549,162	2	9

The amount expended on development and sinking main shafts brought 1,495,550 tons of ore in sight, of which 442,859 tons are in reserve ready for stoping.

The Company originally recovered only 57 per cent. of the value of gold in the ore, this value being arrived at by the addition of the assay value of the residues to the total gold recovered.

This percentage has been increased gradually up to 86 per cent., which is the extraction for the last half of the financial year ending March 31, 1897, when a slimes plant was added to the already existing equipment of mill, cyanide, and concentrating plants.

The Company first got its rock from open trenches, which was a cheap method for a time. During the second year of its existence is started actual underground work. The total costs, exclusive of capital expenditure for that year, averaged £1 13s. 7d. Last year, with the additional cost of three separate secondary treatments, the costs, inclusive of capital expenditure, were £1 8s. 5d. Although the mining costs have shown no great reduction, the milling costs have been reduced from 11s. in 1890 to 3s. in 1896. The position of the property in 1890 was most exhaustively dealt with in a report by myself, which deals fully with the difficulties and imperfections of gold recovery on these fields.

Leaving out of account the first year of the Company's returns, which was abnormal, the yield per ton has been fairly constant, though slightly increasing during the last two years, and varies from £1 15s. 6d. during the second year to £2 7s. 2d. for the last financial year.

THE GELDENHUIS DEEP, LTD.

This Company was formed in January, 1893, with a capital of £350,000 in £1 shares. The vendors received 175,000 shares, while 90,000 shares were issued for working capital, realising £94,500, and 85,000 shares were kept in reserve.

The property consisted of 212 claims, and work was commenced at once.

In 1894 it was found advisable to issue debentures to the amount of £160,000, and in 1895 15,000 out of 85,000 reserve shares were sold, realising £103,474 15s., and further sums were gradually borrowed to complete the development and equipment, on which, altogether, about £410,000 will be spent. This is the cost of putting the mine on a paying basis, and, after all the experience gained in former years, there can be no doubt that the works here, as well as at other subsidiary companies of the Rand mines, are of an extremely high order.

The mistakes of early days have been avoided as far as possible, and every effort has been made to introduce the latest improvements.

The mill started crushing towards the end of 1895, and the early results were poor and unsatisfactory.

During the last three months of that year the yield was only 18s. 2d. per ton, while the costs were 26s.

In 1896 the yield was raised to an average of 27s. 4d., with working costs of 25s., while for April of the current year the working costs, including sorting, were 26s. 5d. per ton, the yield 37s. 3d., and the monthly profit £9,018.

The Geldenhuis Deep is only one of the subsidiary companies of the Rand Mines, Limited, which, as shown by the last annual report, has seven other important companies already in course of development and equipment, and expects to require £3,630,065 to put them on their initial running basis of 710 stamps, with the intention of eventually increasing them to 1,300.

The total nominal capital of the eight companies is £3,607,391, a sum almost identical with the amount of cash estimated to put them on an earning basis.

I trust that I have not wearied you with all the details I have given you concerning the history of these mines ; I have laid everything so fully before you in order to show you what has been the work of one group of capitalists on these fields. I do not for a moment wish to imply that our firm has been the only one to achieve brilliant success, as you have been already, or will be, informed by the representatives of the other houses regarding the good work done by them.

You will have seen from the struggling history of many of these companies that after the early boom of 1889 there was a most serious depression, during which all the mines suffered, but it was during this very depression that the foundations were laid, by means of hard, earnest, and intelligent work, of the revival which followed in the year 1895. We now again are experiencing a period of the most acute depression after the recent boom, but there is an enormous difference, which I cannot too strongly impress upon you, between the position and hopes of the industry at the present moment and during the preceding relapse. In 1890 the industry was still young, it was undeveloped, and there was, as I have endeavoured to show, immense scope for improving mining results, both as regards working costs and extraction of the gold.

Now, in 1897, the class of machinery on these fields can be considered the most perfect of any gold fields in the world, the various processes dealing with the extraction of gold are rapidly approaching practical perfection, and our working costs have been decreased until we can scarcely reduce them further without the Government's help; with this help, however, we can still make great reductions.

I have tried, and I cannot too earnestly try, to impress upon you that the very men who in the early days obtained their profits from rich mines like the Robinson and Ferreira, freely put back this money into other mines, like the City and Suburban, the Henry Nourse, and the New Heriot, then struggling under the greatest difficulties, and after reaping the fruit of their energy and intelligence here, again turned their resources to gigantic enterprises like the Rand Mines.

Before pursuing further the investigations of these fields, I desire in a way to compare the yield, cost, etc., shown in the foregoing statements, with those of other gold mines in the world.

It must be remembered that to make comparisons applicable, the conditions under which work is carried on must be taken into account. Even comparing the twenty-nine different mines in the foregoing list, it will be found that different conditions exist. Some companies sort their ore, and others do not, some work only one reef, others two or three, varying in thickness, hardness, etc.; it is obviously unfair, for instance, to compare the costs of a company exploiting for the most part only one thin reef, and sorting out 40 per cent. of its rock, with a company which is not sorting and working two reefs averaging 10 or 12 ft. in thickness. The scale on which work is conducted must also be taken into consideration, and it is presumable that companies with large stamping power should have an advantage in working costs over those of much smaller power, and companies who are treating ore by two or three secondary processes should be under a disadvantage when comparing costs with those employing only one or two. It will be noticed that the average stamping power of the twenty-nine companies is about eighty-five stamps per company.

COMPARISONS.

I will commence with the Alaska Treadwell Mine, the annual report of which Company for 1896 I beg to put in evidence.

From this report the operating costs are seen to be as follows :

Operating costs on 263,670 tons (all construction charged directly to operating).

	Dollars per ton of ore.	Shillings per ton of ore.
Mining	·5491	2·29
Milling	·3476	1·45
Chlorination	·1138	·47
General Expenses (Douglas Island)	·0819	·34
do. (San Francisco)	·0218	·09
London Office Expense	·0112	·05
Bullion Charges, Freight, Insurance, &c.	·0372	·15
Total Operating Costs	1.1632	4·85
Net profit for year	1·8862	7·86
Total Yield	3·0494	12·71

The wages paid were as follows :—	Per diem. Dols.	Per diem. s. d.
Miners, with board and lodging	2·50	10 5
Labourers "	2·00	8 4
Drillmen, with bonuses and board and lodging (summer)	2·50	10 5
Drillmen, with bonuses and board and lodging (winter)	3·00	12 6
Indians (paid daily)	2·00	8 4

MILLMEN.

	Dollars per month.	Shillings per month.
Concentrators, with board and lodging	65·00 to 100·00	£13 10 0 to £20 16 8
Feeders ., "	70·00 to 100·00	14 11 8 to 20 16 8
Amalgamators " .,	90·00 to 100·00	18 15 0 to 20 16 8

CHLORINATION WORKS.

	Per diem. Dols.	Per diem. s. d.
Roasters, with board and lodging	2·50	10 5
Roasters (helpers) „	2·00	6 4
Floormen „	2·00 & 2·25	8 4 & 9 4½

MACHINE SHOP.

	Per diem. Dols.	Per diem. s. d.
Mechanics, with board and lodging	2·00 to 6·00	8 4 to 25
Blacksmiths „	4·00	16 8
Blacksmiths (helpers) „	2·00	8 4

The Alaska Treadwell Company, however, is situated on an island with a good harbour. The mill is near the ocean, with the tailings running into the water, and is worked by water power for the greater part of the year ; the lode varies in thickness from 50 ft. to 426 ft., and the mining is more or less quarry work. The number of stamps is 240. The mine is most favourably situated for obtaining supplies at low rates, as is shown by detailed account, which I beg the Commission will compare with the rates of similar supplies on these fields, especially dynamite.

The following table is roughly made out to show the relative prices paid for stores at the Alaska Treadwell Mine and at the Crown Reef here for 1896, on the basis of the amount used at the former mine :

Article	Amount used	Alaska price	Crown Reef price
		£ s. d.	£ s. d.
Dynamite	200,089 lbs.	5,134 11 2	17,354 12 2
Fuse	14,314 coils	474 5 0	268 7 9
Caps	75,182	168 5 0	150 8 0
Timber	14,909 cubic ft.	482 0 0	3,168 2 0
Steel, Mining	25,519 lbs.	429 16 5	637 18 0
Oils	6,545 gallons	428 3 5	1,309 0 0
Candles	272 boxes	177 15 0	145 0 0
Mortars	2	96 0 0	304 0 0
Mortar Liners	58,058 lbs.	756 12 0	1,209 10 10
Cam Shafts	3	10 st'ps 59 4 0	5 st'ps 52 10 0
Guide Blocks	120 pair	22 16 0	856 0 0
Shoes and Dies	151,922 lbs.	2,178 17 3	2,278 16 8
Screens	1,300 sq. ft.	109 6 0	97 10 0
Heads	12	49 16 3	78 0 0
Sulphuric Acid	328,000 lbs.	1,000 0 0	4,100 0 0
Salt	455 tons	790 8 6	2,733 0 0
Bar Iron	63,503 lbs.	275 9 6	529 3 10
Lead	1,461 lbs.	18 4 6	52 11 6
		£12,651 10 0	£35,324 10 9

If the same proportion exists for the balance of the stores used by the Alaska Treadwell, which are not classified above, and which amount to further £3,500 exclusive of coal, these total stores, costing £16,100 in Alaska on the above basis of prices, here amount to about £45,100, thus increasing their costs by £29,000, and the total cost per ton at the Alaska Treadwell Mine by 2s. 2d. on the tonnage milled : 263,670 tons.

On the other hand, if the Crown Reef Company had been able to obtain its stores last year at the above prices ruling in Alaska, their supplies, exclusive of coal, which actually cost them £85,100, would only have cost £30,500, which would be a saving of £54,600, or no less than 5s. 6d. per ton, on their tonnage of 198,236 tons.

	Crown Reef	Alaska Treadwell
		1 dol. = 4s.
Tons crushed	198,236	263,670
Pounds of Dynamite used per ton crushed	1·10	0·76
Tons mined and milled, secondary treatment, and general expenses per man per day	0·31	4·14

	Crown Reef		Alaska Treadwell 1 dol. = 4s.	
	Cost per ton milled	Per cent. of total cost	Cost per ton milled	Per cent. of total cost
	£ s. d.		£ s. d.	
Labour, total white and black, including food	0 15 6	57·98	0 2 11	63·78
Coal	0 2 7	9·79	0 0 5	8·57
Dynamite	0 2 5	9·11	0 0 5	8·37
Cyanide Zinc and Royalty	0 1 8	6·35	...	...
Timber	0 0 5	1·55	...	0·79
Steel, Mining	0 0 4	1·32	...	0·72
Oils	0 0 3	0·87	0 0 1	0·92
Candles	0 0 4	1·20	...	0·23
Mill Spares	0 0 5	1·52	0 0 4	6·54
Fuse and Detonators	0 0 2	0·57	...	1·02
Trucks, Wheels, and Rails	0 0 2	0·69	0 0 1	0·01
Pipes and Pipe Fittings	0 0 1	0·36	...	0·50
Sundry Stores, General Expenses Electric Light and Drill Spares	0 2 4	8·79	0 0 2	3·46
Chlorination Supplies	...	...	0 0 3	5·09
Total Cost per Ton	1 6 8	100 %	0 4 8	100 %

This shows that the relative proportion of cost for labour at the Alaska Treadwell is somewhat higher on a percentage basis than at the Crown Reef, and also that the great lowness of cost is, in addition to the cheapness of supplies, due to the fact that the tons mined and milled per man per day are in a ratio of thirteen to one at the Crown Reef.

The next comparison of cost I wish to make is that of the Deadwood Terra Gold Mining Company, Dakota, U.S.A., given me by the manager of the Geldenhuis Deep, who previously to coming out here was manager of this property. It was on account of the remarkably low costs ruling there that we were induced to obtain his services here on the Rand.

The total mining costs of this Company in 1895 are shown to be 1·37 dollars, made up of

Mining	88·539 cents = 3s. 8d.
Milling	48·861 cents = 2s.

The yield of the ore being 1·74 dollars = 7s. 3d. per ton.

If it interests you, further particulars can be obtained from the manager of the Geldenhuis Deep. He informed me that the width of the lode varied from 25 ft. to 75 ft., the deepest shaft was 600 ft., and no secondary treatment was used. The fuel was coal obtained by rail, and ruling rates of wages were as follows :

MILLHANDS.

Engineer	2 at $3·00 per shift of 12 hours.
Foremen . . .	2 „ 2·50 „ „ „ „
Foremen helper	1 „ 2·00 „ „ 10 „
Amalgamators	2 „ 3·50 „ „ 12 „
Amalgamators	4 „ 3·00 „ „ „ „
Feeders	2 „ 2·50 „ „ „ „
Oilers	2 „ 2·50 „ „ „ „
Carpenters	1 „ 3·50 „ „ 10 „
Carpenter helpers	1 „ 2·50 „ „ „ „

Miners in Deadwood Terra Mine received 3·00 dollars per shift of 10 hours and shovellers 2·50 dollars.

In the Homestake and Highland Mines and Mills all of this labour is paid 50 cents per shift more.

The Deadwood Terra Mine ran 160 stamps.

Regarding the Gold Mines of California the total costs in some of the principal mines vary from 10s. to 38s. per ton, depending on local conditions.

Mr. Leggett, a more recent arrival than myself, can give you fuller particulars.

The next comparison of cost I wish to put in evidence is taken from *Mineral Industry*, page 312, in which it states :

" Mount Morgan Mine reports for 1895, the cost of working last year was almost 12 dollars a ton."

Mineral Industry, 1895, same page : " Mysore Company in India treated 60,654 tons and cyanided half the tailings, cost 9·50 dollars per ton." *Mineral Industry*, 1895, page 319 : " Milling in four districts in the U.S.A. is averaged by P. A. Richards as under :

Black Hills	70 cents a ton.
Gilpin	75 „ „
Grass Valley	80 „ „
Amador	46 „ „ (soft ore)."

Taking the average of the first three districts we get 75 cents a ton, or 3s. This is practically the same as the Crown Reef cost for the past two years, including stone crushing.

From this it is seen that the average cost of milling in the Black Hills, Dakota, and the Gilpin Country, Colorado, and Grass Valley and Amador Country, California, is 2s. 10d. per ton, and the mining cost in not given, but varies with local conditions and the width of the lodes.

EL CALLAO

The next comparison is a table of results showing the general operations of the El Callao Company, from its formation up to June, 1894.

Work was first started in 1870 on a small scale, and by people who had no previous experience in mining. The yield per ton is seen to have varied from 5·66 ozs. in 1884 to 0·6 ozs. in 1892, the average for the whole period being 2·03 ozs., or 155s. per ton.

Table showing the General Results of the operations on "EL CALLAO" Lode, since formation of the Company.

PERIODS	Lode area worked on incline. Square metres	Lode Average thickness. Metres	Ore Gross Yield in tons	Gold Gross Yield in ozs.	Gold Yield per ton ozs.	Gold Gross Yield Value £	Gold Yield per ton Value s.	d.	Mining Costs per ton s.	d.	Milling Costs per ton s.	d.	Miscellaneous Costs per ton s.	d.	Total Cost per ton s.	d.	Total Dividends Paid £	s.	d.
1870 to 11th March, 1881	22,102	1·52	91,046	318,855	3·50	1,218,115	267	7	...		...		...		...		313,433	13	5
11th Mar. 1881 to 31st Dec., 1883	15,461	1·60	67,073	300,650	4·48	1,148,700	342	6	...		...		...		149	2	497,886	4	11
1884	7,513	1·54	31,261	177,055	5·66	677,569	433	5	86	7	29	10	5	10	122	3	383,300	15	0
1885	8,949	1·94	46,868	114,500	2·44	435,040	185	8	59	1	20	4	4	5	83	10	181,429	0	5
1886	13,867	2·00	74,399	181,300	2·40	685,860	184	4	43	9	14	8	1	5	59	10	436,962	17	1
1887	13,273	1·75	64,215	73,872	1.15	282,000	87	10	45	10	7	3	1	10	54	11	58,772	15	7
1888	13,528	1·45	54,438	52,598	0·97	199,994	73	5	51	9	7	11	2	0	61	8	5,110	13	7
1889	9,755	1·68	56,389	52,973	0·93	204,134	72	4	42	5	6	3	5	2	53	10	20,442	14	1
1890	13,113	1·52	53,977	49,432	0·93	189,829	70	3	46	8	6	3	1	9	54	8	20,442	14	1
1891	16,321	1·33	59,284	34,774	0·59	132,270	44	7	32	8	5	0	2	1	39	9	...		
1892	13,825	1·40	52,823	31,931	0·60	120,297	45	6	33	3	5	0	3	1	41	4	...		
1893	11,000	1·36	49,085	34,537	0·86	131,559	65	8	32	10	6	3	3	0	46	1	...		
1st January to 30th June 1894	3,048	1·40	11,607	8,417	0·73	32,063	55	2	45	9	6	3	1	8	53	8	12,697	6	8
Totals	161,755	1·54	703,465	1,430,894	2·03	5,457,432	155	2									1,940,478	14	10

The working expenses varied from 149s. 2d. per ton, which was the average for the first thirteen years, to the minimum of 39s. 9d., in 1891, the total average being 53s. 8d. for the whole period.

The total gross yield from this mine is shown at £5,457,432, while the dividends declared have only amounted to £1,940,478, in spite of the phenomenal richness of the ore.

This, with the exception of the Mount Morgan Mine, can be considered to have been at one time the richest gold mine in the world.

The success of this mine induced capital to be invested freely in other lodes, which, however, were not of the same richness, though in many cases of as high a grade as that ruling on these fields, and thirty-one companies were started, which erected 758 stamps. At the present time the industry is almost dead, there are only ten stamps running on these fields, and these are working without a profit, and the celebrated El Callao Mine has just closed down in debt to the extent of 150,000 dollars. Let us now enquire into the above high costs and final languishing of the fields. In the early history the discoverers worked on a very small scale with very little experience, and their immense costs are not to be wondered at considering the general political and climatic conditions. The Callao Mine is situated about 150 miles south of the Orinoco River, in a latitude of about 7 degrees north of the equator, without railway communication, and with very bad roads, and in its early history was considered most unhealthy, so that high wages had to be offered to induce skilled men to come to the property.

In 1884 a new regime was started, and high grade machinery with increased stamping power and high grade men put to work, the Company giving them a free hand and every encouragement to do their best, and expenses were lowered in eight years from £6 2s. 3d. to £1 19s. 9d. per ton. The cardinal feature in this reduction was the improvement in machinery and mining methods, but another was the encouragement of negro labour obtained from the West Indian Islands. At first this class of labour was considered hopelessly incompetent, but by patient training and judicious graduation of wages in proportion to work done it was finally possible to run the mine with 11½ per cent. of the white men originally required, and the blacks were better able to stand the climate.

The Government of Venezuela, which was not in sympathy with the alien mining population, believed in high and onerous tariffs, monopolies and concessions, and did very little to foster the industry ; in fact, tried in every way to extort as much as possible out of it.

The present unfortunate condition of the mining industry there is, I think, in no small measure due to the attitude of the Government. The ruling rates of wages during late years, when good work has been done, are :

White, average per month	£35
(This does not include the Superintendent)	
Blacks, per day	6s. 6d.

I would state in this connection that Mr. H. C. Perkins, late General Manager of the Rand Mines, initiated in 1884 the better working and equipment of this mine. I succeeded him as Superintendent in the latter part of 1887, and remained until the middle of 1889. Mr. Webber, now General Manager of the Rand Mines, succeeded me, and remained until he came here in 1891, and he in turn was succeeded by Mr. Searle, now Manager of the Crown Deep, who remained until 1896. Mr. Searle, the most recent arrival from El Callao, will be best able to give you further details regarding the working of the mines and the Government of the country, should you desire them.

Although now out of date as a treatise on the comparison of working costs, I wish to put in evidence an interesting little pamphlet published in 1886 by Mr. Hamilton Smith, who deals with the relative costs in the United States and Venezuela Gold Fields, and gives the relative conditions under which work is accomplished.

The table on page 10, which I herewith incorporate in my statements, gives the substance of the comparisons, and is as follows :

MINE	Period	Average No. of Stamps	Tons crushed in one year	Average No. of tons crushed by each stamp per month	Costs of milling per ton in dollars	Total costs per ton in dollars	Costs of milling per ton in shillings	Total costs per ton in shillings
Sierra Buttes . .	1885	76½	54,493	59	·56	5·83	2·33	24·25
Plumas Eureka .	1885	60	55,973	78	·61	5·57	2·54	23·17
Homestake	1882-3	200	170,074	75	1·17	4·03	4·87	16·76
,, . .	1883-4	200	191,505	80	1·21	4·19	5·03	17·43
,, . .	1884-5	200	213,190	89	1·01	3·25	4·20	13·52
Father De Smet	1883	100	104,100	85	...	2·49	...	10·36
,, .	1885	100	106,855	89	...	2·12	...	8·82
Caledonia .	1885-6	...	48,848	...	·88	2·95	3·66	12·27
El Callao . .	1882	60	22,405	31	11·19	45·34	46·55	188·61
,, .	1883	60	24,750	34	...	44·33	...	184·41
,, . .	1884	60	30,936	43	7·25	35·17	30·16	146·31
,, . .	1885	80	47,223	49	4·98	21·96	20·72	91·35
,, new mill {	May, 1886	40	...	83	about 3	about 15	12·50	62·50
New Potosi {	11 mos 1884	25½	7,456	27	...	46·96	...	195·35

I regret that I cannot put in more details of the working costs in other countries compared with the elaborate details given by me for these fields, although I have gone to considerable trouble to obtain further figures, but this is in accordance with the statement already made by me that information is more generously given here than in any other mining district in the world.

From the figures, however, that I have given it is evident that our average cost of 3s. 8d. for milling (some companies running under 3s.) compares most favourably with the milling costs of other mines in the world, working under similar conditions, and speaks eloquently for the excellence of our machinery and mill organisation, when the high price of labour and supplies is taken into consideration, and when we have to conserve and pump our mill water. These milling costs can still further be reduced, but only in items of labour and supplies. To show the improvement in milling on the Rand since 1890, I would state that I had cause to examine a mill of 100 stamps in that year, at which the milling costs were 10s. 2½d. per ton, not including cost of shoes and dies, the number of white men employed being twenty-nine, and the number of black 188.

The number of men employed to-day at the Geldenhuis Deep Mill (155 stamps) is about twenty white and about twenty black, including engine-men.

Cyanide costs of these fields are not comparable with any others at my disposal, but may safely be assumed to be the lowest in the world. The main costs are seen to be in mining, including mine development, the average amounting for the twenty-nine companies to 17s. 7·12d. per ton. This is abnormally high, and is the department in which our principal future reduction is to be made. This department has little to expect from improved machinery, and the main hope of reduction lies in increasing the efficiency of labour or decreasing its wage, or both, and also in decreasing the cost of all supplies, especially dynamite and coal.

Before leaving this subject I wish to state that a mere reduction in working costs is not the only thing to be aimed at, as I believe there is still considerable scope for raising the yield by extending and improving the system of sorting in vogue at some of the mines, and mining underground a minimum amount of waste. In this way an apparently higher cost per ton on the basis of the tonnage milled might be shown while actually cheaper work was being done, but larger profits would result. Mr. Johns, who is the pioneer of this system on these fields, can give you better information about it than I, though I also have been, and am, introducing it at all the mines with which I am connected. I roughly estimate that now, for the twenty-nine principal mines, not over 8 per cent. of the ore is sorted, and I think in time this figure will be more than doubled. The only further improvement I see possible from an engineering point of view is the introduction of slimes treatment throughout the mines.

This will be an expensive matter with the freight rates, etc., ruling here, and will probably cost for plant approximately between £120 and £200 per stamp, according to the size of the mill. The increase of yield and profit will depend on the grade of the slimes leaving the mill, which varies in the case of every mine, and in some cases may not warrant the expenditure of erecting such a plant.

I have included in this portion of my statement an immense amount of figures which have been compiled with great care, but owing to the short space of time which I have had at my disposal in drawing this up, it is possible that a few small errors may have occurred, and if so I hope I shall be informed of the fact by the gentlemen on the Commission, as I do not wish a few possible clerical errors to vitiate any of the arguments I am advancing.

LABOUR.

I have already explained that the engineer is about at the end of his tether as regards further improvements under present conditions.

From the analysis of working costs on labour and supply basis, it is evident that labour is the most vital point of all, and the point towards which our chief attention must be directed.

The summary of my statement for six companies shows that labour figures on a basis of total costs per ton as follows :

White labour .	.	8s. 7·78d. per ton	30·18 per cent.
Black „	.	6s. 9·62d. „	23·73 „
Kaffir food	.	1s. 2·24d. „	4·14 „
		16s. 7·64d. per ton	58·05 per cent.

The State Mining Engineer for the year 1895 gives

White labour at	.	.	34·3 per cent.
Black „	. .	.	28·6 „
			62·9 per cent.

Taking the mean of these two estimates, we have, roughly speaking, 60 per cent. of our working cost appearing under the item of labour.

Now, what have we to face in our labour problem? It is, first, that our labour is accomplished in the proportion of, roughly, one white man to eight or ten black. Some of the white labour is the best that money can command, and is culled from all over the world It is very highly paid when compared to labour in old-established countries, where climatic and general conditions are favourable, but when compared to new fields, to which men only go for high wages, it is not excessive, as shown by the statement made by Mr. Seymour with regard to the price of labour in Nevada, Montana, and British Columbia, and also by the figures I have myself already given of wages in other districts.

I now give a tabulated statement showing the average daily wages paid to the employés in the older mining districts of various countries :

Country.	Mines.	Average, Surface.	Average, Underground.	Reference.
Belgium. . .	Coal	2s. 3·8d. to 2s. 11·7d.		Engg., v. 20, 1888-9-90
Bilbao, Spain. .	Iron	1s. 6d. to 2s. 6d.	2s. 6d. to 3s. 4d.	Engg., v. 44, p. 271
Durham, England .	Coal	2s. 6d.	3s. 8d.	Engg., v. 53, p. 413
France . . .	Coal	2s. 4d. to 2s. 10¼d.	3s. 1½d. to 3s. 11d.	Engg., v. 53, p. 413
Germany .	Coal	2s. 2d. to 3s.	2s. 4d. to 3s. 3d.	Engg., v. 53, p. 413
Hungary . .	Coal	0s. 10d. to 1s. 2d.	2s. 0d. to 2s. 6d.	Engg., v. 54, p. 179
United States :				
California .	Gold	4s. 0d. to 8s. 0d.	10s. 0d. to 12s. 0d.	Approximation.
Georgia .	Coal	3s. 10d. to 4s. 0d.	6s. 0d.	Engg., v. 52, p. 582
Pennsylvania .	Coal	4s. 9d.	5s. 0d.	Engg., v. 53, p. 413

NOTE.—In Hungary and Spain the number of hours per week is, in summer, as high as seventy-two ; while in the other named countries the hours vary from fifty-four to sixty-six. In the American coal mines the average number of days worked in 1894 was only 210, but wages are based on the actual weekly payments, divided by six.

Mr. Goldmann's statement regarding 3,620 miners on the fifty-three companies, Witwatersrand gold fields, is food for serious reflection, showing that 54 per cent. of the men are single, 33 per cent. are married, with their families in other countries, and the small balance of only 13 per cent. consists of married men with their families here.

What does this mean?

It means that the majority of the men have come out here simply attracted by the high wages, and have not deemed it advisable, owing to the condition of the country, to make it their home.

Other witnesses have shown how wise they were in so doing, owing to the great cost of living prevailing here.

What would it mean if we materially reduced these men's wages? Would it not be that the best men for the industry and the Republic, and those whom we are most anxious to keep, would leave, and we would be left with the improvident, and those who could not command work in their own country?

Now, regarding the native labour, which comprises in numbers by far the greater proportion of labour we are using. What has been the keynote of our trouble? Lack of supply in proportion to the demand, and inefficiency and ignorance of this class of labour, which is not trained to the intricate work demanded of it.

Far more skill is required of the Kafir on the Witwatersrand than is the case for the most part on the diamond fields or in any agricultural pursuit in South Africa. The boys come here raw, some very young, often with weak physiques, and are all comprised in the same classification. They are accustomed to their own simple ways, and desire to return to them as soon as possible. They come, in fact, only in order to make enough money to return to their kraals with sufficient means to enable them to marry and live in indolence.

There is much latent possibility in them for learning, but they leave us often as soon as they become really useful; and by the various companies vieing with each other to obtain their services they have become masters of the labour situation.

If they had facilities for making their homes in this country, and if they could be induced to remain with us, I am satisfied that their efficiency could be increased twofold, and they could even be trained to do much higher grade work than they are now employed at.

This fact is illustrated in these fields by the boys who have worked long periods being able to finish their task in half the time that raw boys require, and this illustration is further strengthened by the well-known evolution of the negro in America since the abolition of slavery, and also by my own experience in dealing with the West Indian negro in Venezuela.

The Pass Law and Liquor Law have been modified and strengthened, on the petitions of the industry, to give more control to the mining companies in dealing with this class of labour.

These laws, though not perfect, are good and useful if well administered.

But what is the testimony already brought before you in this regard ?

Witness after witness has testified to the unsatisfactory manner in which they have been carried out, and the elaborate tabulated statement, brought in evidence of the seventy-four companies, is eloquent on the subject of the Liquor Law, whose maladministration is a great source of loss to the companies, danger of life to the natives, and discredit to the Republic.

If you should desire further evidence on this subject, I would ask you to call Mr. A. Grant, manager of the Nigel Mine, who can give, I believe, even stronger testimony than any before you.

Regarding the Pass Law, there has been, as far as I am aware, no witness yet before the Commission who has stated that this law, as administered, had benefitted his company, and Mr. Goldmann has informed you that out of thirty-three companies employing 19,000 boys monthly, 14,000 have deserted since the new Pass Law came into operation, without one single one of these deserters having been brought back to justice.

In my opinion, the Pass Law, though good as a temporary expedient, is only the kindergarten of the native question, and before these fields can ever reach their maximum possibilities, the whole question of native labour must be dealt with on broader and more liberal lines, and modifications and suggestions in regard to it must be constantly expected by the Government.

The medium and lower grades of white labour here have to a great extent been demoralised by the black labour ; and although there are exceptionally energetic and earnest white workers here, there is a lot of indolence and incapacity shown by many. The majority, although they may have skill in doing work themselves, lack the faculty or interest in getting the best work out of the black labourer.

What we require from both black and white labour is greater efficiency, which, if really obtained, renders rate of wages a secondary consideration, as shown by the bonus system in sinking deep level shafts. Of course, we desire to get the unit of wage as low as is consistent with the contentment of the labourer.

What the management of the mines must aim at is to encourage in every way the efficient man, and give him every preference and advantage over the inefficient man, and to elevate the quality of native labour, which at the same time will justify a greater rate of wages being paid to the whites. Contracts, piece work, bonus systems, and uniformity in accounts should be encouraged, and the companies should not strive to make statistical records of low wage rates, but rather accomplish cheap results judged by work actually done.

The Government on its part should endeavour to lessen the good workers' living expenses, and make them interested in the welfare of this country, so that they will be satisfied to remain here, and be contented with a lower wage.

The Government should also do everything to bring about an abundant supply of black labour, and give the mines reasonable control of it through the proper administration of the Pass Law and all other laws connected with the native question, and should encourage this labour to the utmost extent, realising its vast importance to the prosperity of these fields and to this Republic.

DYNAMITE.

The next item figuring on our expenditure sheet is dynamite, 2s. 10·3d. per ton, and 9·92 per cent. of the total cost.

So much evidence has been already brought up by various witnesses on this subject that there is very little further to be said as regards the price at which dynamite can be landed here free of duty.

I, however, am in a position to state that in February last I had an offer from reliable people in America of dynamite (70 per cent. nitro-glycerine) to be delivered at any port in South Africa at 17 cents per lb., or 35s. 6d. per case of 50 lbs., in large quantities. With landing charges, agents' fees, railage, colonial duty, etc., the price would be raised to 42s. 7d. per case, delivered in Johannesburg free of Transvaal duty.

This price would practically save 50 per cent. on the present price of dynamite, No. 1, or 1s. 5d. per ton on the basis of the tonnage milled.

There is, however, a great deal more to be said on the subject of the Dynamite Concession than mere £ s. d., and I believe I am the first witness that has dealt with the subject from this standpoint ; and, after all, it is the main standpoint, as this concession, yea, and the principle of concessions, is one of the fundamental causes which have brought about the estrangement between the original population of this country and the Uitlanders.

Take the Government Volksraad Dynamite Commission's Report on the subject of dynamite, and it is there stated that the main articles required for dynamite manufacture are not found in this country, and that the concessionaires have not adhered to the terms of their contract. What justification is there, then, in fostering such a manufactory, especially in the early struggling stages of the mines, which are the main source of revenue to the whole country, and the essence of its prosperity?

Take the concessions as a whole, what justification, intrinsically, is there for them at all? What nations are the most prosperous, and which of them have concessions?

I speak feelingly and knowingly, as an American. Although we have protective tariffs to foster industries, we have the most complete internal system of free trade. There is no such thing known as a trade concession in the United States of America.

What is there vitally wrong about a concession? It seems to me the crux of the whole thing is that it places the production of an article of necessity in the hands of a few, who are given opportunity to profit by the necessities of the many. The many rebel against this principle, and the Government reaps inadequately, and derives little benefit from the burden it imposes on its people. Vexations crop up on all sides; excuses from managers and directors as regards working costs; excuses from directors to shareholders, and general ill-feeling and destruction of confidence; these are the results of concessions.

In order to keep their concessions, concessionaires are tempted to use every specious device of argument to the Government to continue their profits. It gives opportunity to unscrupulous persons to play upon the best and worst motives of your people, and it is wrong in principle and in practice.

This general line of argument can be applied to all the concessions in your country.

The Chairman's patriotic suggestion to manufacture machinery, etc., in this country, and thus to give employment to a far greater number of people, is, I think, a most dangerous one, though made with the best intention. It might be taken advantage of, and new burdens in the shape of concessions would then be imposed upon us.

There is no doubt that if our population were such that we were ripe for the manufacture of machinery it would be an excellent thing, but to produce at enormous cost the raw materials and to manufacture all the complex machinery now required, would take almost a larger number of white men than is now actually employed at the mines, and the initial stages of such an industry could only be rendered a sufficient inducement to capital by heavy protective duties, and consequently immense further taxation of the mines.

Let us first get our full growth before such a thing is earnestly considered, and then also allow every form of healthy competition.

RAILROADS.

Referring to the table of analysis which we have already dealt with, we find that we have dealt with about 68 per cent. of the mines' total costs. General charges figure at 5·22 per cent., which cannot well be further detailed, so that we have then 26·81 per cent. made up of supplies, exclusive of dynamite, remaining.

It is, then, in this item of 26·81 per cent. of our total costs, as well as on the 4·14 per cent. for Kaffir food, already included under the heading of labour, that the railway charges have the most direct bearing. Previous witnesses have gone into great detail with regard to this railway question, giving all manner of comparative figures and suggestions, besides dealing with capitalisation, profits, etc. Mr. Fitzpatrick has shown you that if the English rates of $\frac{1}{2}d$. per ton per mile for coal had prevailed on the Netherlands line, 5s. 8$\frac{1}{2}d$. per ton of coal, or about 1s. of our total costs per ton of ore milled, would have been saved; while Mr. Seymour has shown that by reducing the present rates per ton per mile to 1$\frac{3}{4}d$. without terminal charges—which rate would still be five times higher than in America—and by employing side discharge trucks and so doing away with bagging, a yearly saving to the industry of £407,500 would be effected.

I would emphasize that this saving would be on coal alone; and if the rate per ton per mile for all other goods were reduced to the basis of other countries, and the Government used its influence with the other South African railroads to effect a similar general reduction, our supplies (notably timber) could be obtained at

R

an enormously cheaper rate, the reduction varying, as shown by Mr. Albu and other witnesses, from 20 per cent. to 40 per cent.

I see no new facts that I can bring before the Commission in this matter, and I will now simply touch upon it from the standpoint of an engineer, and take the general rates of transport per mile as given by Mr. Seymour, selecting machinery for the purposes of comparison, which is the most favourable for the Netherlands Railway. This shows :

COST OF TRANSPORT BY RAIL IN PENCE.

			Ratio.
Machinery—American	0·51 pence per ton per mile	.	1·000
Do. Cape Railways . .	2·34 do.	.	4·565
Do. O. F. S. Railway . .	2·34 do.	.	4·565
Do. Natal Railways . .	3·04 do.	.	5·931
Do. Portuguese . . .	4·07 do.	.	7·940
Do. Netherlands (Cape) . .	7·69 do.	.	15·000
Do. Netherlands (Natal) .	5·06 do.	.	9·871
Do. Netherlands (Delagoa) .	4·27 do.	.	8·330
Do. English Railways . .	1·12 do.	.	2·190

I find from Mr. Dawsey's " *Comparison of English and American Railways,*" and the Chamber of Mines Report, the following information, which I tabulate :

COST TO BUILD RAILWAYS (OPEN FOR TRAFFIC).

		£ s. d.	
1883—England	. Cost per mile,	41,846 0 0	Standard gauge.
1883—America	Do.	12,756 0 0	Do.
1895—Cape	Do.	9,056 9 1	Narrow gauge.
1895—Natal	Do.	15,254 17 9	Do.
1895—Netherlands	Do.	15,359 6 10	Do.
1896—Cape	Do.	9,406 15 0	Do.
1897—Orange Free State .	Do.	7,479 4 6	Do.

From this it is seen that the Netherlands Railway on their portion of the Cape Line charge fifteen times as much per ton per mile as the American railroads, while their cost of equipment is only in the ratio of £15,359 per mile to £12,756 in America, or a ratio of 1 in America to 1·247 on the Netherlands Line.

I grant, as I have shown for our operating costs in mines, that it is more expensive to work in this country than in America, but nothing like in the ratio of 15 to 1. Why, then, is it permitted to keep up this ratio ?

If the Government has no control of the detailed working of the railway while the Netherlands Company has the management, by all means let it exercise its right to take the railway out of the Company's hands, and run it more in accordance with the rates in other civilised countries.

After all, a fair interest on the road-bed basis open for traffic, after deducting working costs, is, in a case like this, the right standard for fixing its tariff.

The cost of transport by rail ought to be moderate in the Transvaal in view of the absence of heavy gradients, which would cause extra expenditure for tunnels, fills, etc., and of the lines having been laid to follow the natural levels and contour of the country, so that the length of railroads between two stations is often considerably greater than the direct distance between them in a straight line.

These high rates are not only bad for the mining industry, but also for agriculture. The railroads in this country should form a potent factor in its development, by enabling the farmer to sell his produce in the market in competition with goods from other parts of South Africa, and if railroads and agriculture were both on a sound basis here, we should no longer see such an anomaly as we now witness :

America, 10,000 miles away, supplying Johannesburg and the mines with mealies !

Australia, Sweden and Switzerland sending butter and tinned milk, and even California supplying a portion of the preserved fruit !

In this connection, I beg to put in evidence a table taken from the *Statistician and Economist*, 1893-94, showing the relative value in 1892 of the different products of the State of California, where, as you know, gold mining, which started in 1849, was the pioneer industry :

	Products	Value in Dollars	£
1.	Wheat	26,626,584	5,325,317
2.	Gold	9,361,486	1,872,297
3.	Wool	7,260,000	1,652,000
4.	Grapes	4,844,331	968,866
5.	Mealies	1,208,213	241,643
6.	Oats	794,956	158,991

GOLD THEFTS.

There is no doubt that the mines are suffering very serious loss in this direction, and that it is the imperative duty of the Government to aid in putting a stop to it.

The Gold Theft Laws should be amended so that there is no possibility of escaping justice through a mere technical quibble, and the police and detective department should be made more efficient; for it is certainly most discouraging that the Company which went to such expense and trouble in attempting to bring the insidious tempters of their tried and trusted men to justice, should only reap as a fruit of its efforts the better advertisement of the business of illicit gold amalgam buying, illustrated by the escape of Hart, the lenient sentence of Hildebrandt, and the sympathy of a certain class said by the newspapers to be circulating a petition for Hildebrandt's release.

TAXATION.

Several witnesses have most ably gone into this matter, and shown the peril of the Government and mining industry in this connection. It is a recognised fact that the mining industry is the chief source of wealth of the country, and through its development has made great demands upon the Government for facilities to work to the best advantage. It is therefore the duty of the industry to take upon itself the burden of taxation in proportion to the demands it makes upon the Government, and if there are a large number of mines working, this burden will be felt less than if only a few are kept running.

I have endeavoured to show the vast natural resources of these goldfields, and how they differ from others in the world.

A great impetus has been given to mining here through the success of some companies, and in their wake have followed all manner of new mining enterprises; money has been forthcoming on the supposition that this was a permanent industry which would be encouraged and fostered, and that with time a continually lower grade of ore would be payable. I believe that, although the limit for further mechanical and metallurgical improvements is very narrow, there is still immense scope for the management of the mines and for the Government to reduce working costs through the medium of labour and supplies, and if the mining community and the Government work energetically and harmoniously together, I see no reason why in course of time the present cost should not be reduced by one-third, or about 10s. on the average basis of 29s. per ton. But this cannot be accomplished at once. It will take time and earnest efforts on both sides.

If we show that we are effecting this reduction, and that the Government is earnestly helping us, capital will again flow into the country, new mines will be opened up and old ones restarted, and the revenue of the Government maintained and increased by renewed prosperity.

But, if nothing is done by the Government, the comparatively few present working mines must directly or indirectly pay the whole, or nearly the whole, taxation of this country. The direct taxation is certainly small, but the indirect is extremely heavy; and it is obvious that these few mines are utterly unable to meet the enormous strain of supplying the Government with the necessary funds for its yearly expenditure, which in 1896 reached, according to the Budget returns, the huge total of £4,500,000. The obvious result of this condition of affairs will be a deficit in the Government Budget, and the strangulation of the mines.

The evil day may be averted by loans, which the Government can only finance upon the assumed prosperity of the industry ; and if the life is crushed out of the industry by oppressive direct or indirect taxation, it will be harder and harder for the Government to continue raising loans.

If the industry has made mistakes in being over confident and launching out in a greater measure than was intrinsically justifiable, the Government has profited indirectly by this wrong and is party to it.

Now, this line of argument brings me to the main problem we both have to face, *i.e.*

LACK OF CONFIDENCE.

You may wish to call this a sentimental grievance, but it is to my mind the most vital with which we have to deal.

It would appear to a great many who have read the statements in the Press, memorials of grievances and their method of acceptance and treatment, that you do not believe in us, that we do not believe in you or each other, and I fear the world will soon not believe in any of us, if the existing state of affairs continues.

It is no use for the industry and the Government to incriminate each other ; this will only make matters worse.

Granted, on the one hand, that there has been over-speculation on these fields, surely, on the other hand, the Government has been a partner to it by receiving emoluments through fictitious values. It has made boom budgets if we have made boom estimates ; but I do not think there is any necessity to grant this to the world.

What I think the Government should do is to justify the policy of its Republic and its main industry to the world ; to show to the world that we have intrinsic merits here ; that we have the greatest gold fields in the world here ; that there are most earnest workers here ; that the world has been given most exhaustive and accurate statements by the State department and by the industry ; that a great deal of the troubles that we are subject to is made by the world, which has taken our good work as a basis for unjustifiable speculation, and made gambling tables of our mines.

This argument we can bring to the world only in one way, and that is by being united ourselves. How can we be united ?

Let the Government commence by abolishing the Dynamite Concession, taking over the railroads and reducing their rates. This must, however, only be considered as an initial measure and an earnest of the Government's desire to conduct the Government of this Republic on true broad Republican principles.

What have I shown to be the main factor for us to deal with ? Is it not the labour question, involving all the vital principles of Republican Government ?

We have not complained so much against the laws of this country as against their administration. The Government must take us into its confidence. It must allow our trained ability to bear upon the serious problems before us. We must be made partly responsible for the administration of this Government, and to be made responsible we must have representation. I would be a coward not to face this issue, with my bias towards you and Republicanism. I see no other way out of it.

If we are only granted the reliefs prayed for now, we will come back for others shortly, and there will be il feeling and heart-burning as of yore. Face the difficulty fairly and squarely. This country has now an opportunity in its history of showing its true greatness, by giving freely what could not be forced from it. I have not been mealy-mouthed, and I leave it to the Government to devise how it can most safely grant us a voice in its affairs. I think the workers on these fields care nothing for the shaping of the foreign policy or the general government of this Republic. They do not wish to deal with problems in districts in which they do not work or are not interested. What they want is representation and a voice in things which concern the economic problems that they have to deal with.

How to do this I will not be so presumptuous as to suggest ; all I ask is that the present burghers admit their own limitations, lack of experience, and training for these problems, and ask the best and wisest of all countries to aid them.

Let them first believe in their hearts the necessity and right of this demand, and the way will be made clear.

This statement is made by me as an individual, and one in sympathy with Republican principles, and not as one representing any body of men or corporation of this place.

Printed in the United States
By Bookmasters